# Wanderings in Sou

Charles Waterton

Alpha Editions

This edition published in 2024

ISBN 9789362998613

Design and Setting By
**Alpha Editions**
www.alphaedis.com
Email - info@alphaedis.com

As per information held with us this book is in Public Domain.
This book is a reproduction of an important historical work.
Alpha Editions uses the best technology to reproduce historical work
in the same manner it was first published to preserve its original nature.
Any marks or number seen are left intentionally to preserve.

# Contents

| | |
|---|---|
| INTRODUCTION. | - 1 - |
| First Journey. | - 17 - |
| LETTER TO THE PORTUGUESE COMMANDER. | - 55 - |
| REMARKS. | - 56 - |
| SECOND JOURNEY. | - 60 - |
| NOTES. | - 94 - |

# INTRODUCTION.

Plutarch, the most famous biographer of ancient times, is of opinion that the uses of telling the history of the men of past ages are to teach wisdom, and to show us by their example how best to spend life. His method is to relate the history of a Greek statesman or soldier, then the history of a Roman whose opportunities of fame resembled those of the Greek, and finally to compare the two. He points out how in the same straits the one hero had shown wisdom, the other imprudence; and that he who had on one occasion fallen short of greatness had on another displayed the highest degree of manly virtue or of genius. If Plutarch's method of teaching should ever be followed by an English biographer, he will surely place side by side and compare two English naturalists, Gilbert White and Charles Waterton. White was a clergyman of the Church of England, educated at Oxford. Waterton was a Roman Catholic country gentleman, who received his education in a Jesuit college. White spent his life in the south of England, and never travelled. Waterton lived in the north of England, and spent more than ten years in the Forests of Guiana. With all these points of difference, the two naturalists were men of the same kind, and whose lives both teach the same lesson. They are examples to show that if a man will but look carefully round him in the country his every-day walk may supply him with an enjoyment costing nothing, but surpassed by none which wealth can procure; with food for reflection however long he may live; with problems of which it will be an endless pleasure to attempt the solution; with a spectacle of Infinite Wisdom which will fill his mind with awe and with a constantly increasing assurance of Infinite Goodness, which will do much to help him in all the trials of life. He who lives in the country and has the love of outdoor natural history in his heart, will never be lonely and never dull. Waterton himself thought that this love of natural history must be inborn and could not be acquired. If this be so, they ought indeed to be thankful who possess so happy a gift. Even if Waterton's opinion be not absolutely true, it is at least certain that the taste for outdoor observation can only be acquired in the field, and that this acquisition is rarely made after the period of boyhood. How important, then, to excite the attention of children in the country to the sights around them. A few will remain apathetic, the tastes of some will lie in other directions, but the time will not be lost, for some will certainly take to natural history, and will have happiness from it throughout life. No study is more likely to confirm them in that content of which a favourite poet of Waterton's truly says:—

"Content is wealth, the riches of the mind,
And happy he who can that treasure find."

Gilbert White and Charles Waterton are pre-eminent among English naturalists for their complete devotion to the study; both excelled as observers, and the writings of both combine the interest of exact outdoor observation with the charm of good literature. Waterton was born on June 3rd, 1782, at Walton Hall, in the West Riding of Yorkshire, a place which had for several centuries been the seat of his family. His father, Thomas Waterton, was a squire, fond of fox-hunting, but with other tastes, well read in literature, and delighting in the observation of the ways of birds and beasts. His grandfather, whose grave is beneath the most northern of a row of old elm trees in the park, was imprisoned in York on account of his known attachment to the cause of the Young Pretender. As he meant to join the rebel forces, the imprisonment probably saved his own life and prevented the ruin of the family. In his grandson's old age, when another white-haired Yorkshire squire was dining at Walton Hall, I remember that Waterton and he reminded one another that their grandfathers had planned to march together to Prince Charley, and that they themselves, so differently are the rights of kings regarded at different ages, when schoolboys together, had gone a-bird's-nesting on a day, in 1793, set apart for mourning for the decapitation of Louis XVI. Waterton has himself told the history of his earlier ancestors in an autobiography which he wrote in 1837:—

"The poet tells us, that the good qualities of man and of cattle descend to their offspring. *Fortes creantur fortibus et bonis.*' If this holds good, I ought to be pretty well off, as far as breeding goes; for, on the father's side, I come in a direct line from Sir Thomas More, through my grandmother; whilst by the mother's side I am akin to the Bedingfelds of Oxburgh, to the Charltons of Hazelside, and to the Swinburnes of Capheaton. My family has been at Walton Hall for some centuries. It emigrated into Yorkshire from Waterton, in the island of Axeholme in Lincolnshire, where it had been for a very long time. Indeed, I dare say I could trace it up to Father Adam, if my progenitors had only been as careful in preserving family records as the Arabs are in recording the pedigree of their horses; for I do most firmly believe that we are all descended from Adam and his wife Eve, notwithstanding what certain self-sufficient philosophers may have advanced to the contrary. Old Matt Prior had probably an opportunity of laying his hands on family papers of the same purport as those which I have not been able to find; for he positively informs us that Adam and Eve were his ancestors:—

'Gentlemen, here, by your leave,
Lie the bones of Matthew Prior,

>A son of Adam and of Eve:
>   Can Bourbon or Nassau go higher?'

Depend upon it, the man under Afric's burning zone, and he from the frozen regions of the North, have both come from the same stem. Their difference in colour and in feature may be traced to this: viz., that the first has had too much, and the second too little, sun.

"In remote times, some of my ancestors were sufficiently notorious to have had their names handed down to posterity. They fought at Cressy, and at Agincourt, and at Marston Moor. Sir Robert Waterton was Governor of Pontefract Castle, and had charge of King Richard II. Sir Hugh Waterton was executor to his Sovereign's will, and guardian to his daughters. Another ancestor was sent into France by the King, with orders to contract a royal marriage. He was allowed thirteen shillings a day for his trouble and travelling expenses. Another was Lord Chancellor of England, and preferred to lose his head rather than sacrifice his conscience."

Waterton's childhood was spent at Walton Hall, and in his old age he used sometimes to recall the songs of his nurses. "One of them," he said, "is the only poem in which the owl is pitied. She sang it to the tune of 'Cease, rude Boreas, blustering railer,' and the words are affecting:—

>'Once I was a monarch's daughter,
>   And sat on a lady's knee;
>But am now a nightly rover,
>   Banished to the ivy tree.
>
>'Crying, Hoo, hoo, hoo, hoo, hoo, hoo,
>   Hoo, hoo, my feet are cold!
>Pity me, for here you see me
>   Persecuted, poor, and old.'"

He was already proficient in bird's-nesting when, in 1792, he was sent to a school kept by a Roman Catholic priest, the Reverend Arthur Storey, at Tudhoe, then a small village, five miles from Durham. Three years before his death he wrote an account of his schooldays, which is printed in the Life prefixed to Messrs. Warne's edition of his "Natural History Essays." The honourable character of the schoolmaster, and the simple, adventurous disposition of his pupil, are vividly depicted in this account. The following quotations from it show that preparatory schools were less luxurious in the last century than they commonly are at the present day:—

"But now let me enter into the minutiæ of Tudhoe School. Mr. Storey had two wigs, one of which was of a flaxen colour, without powder, and had only one lower row of curls. The other had two rows, and was exceedingly well powdered. When he appeared in the schoolroom with this last wig on,

I know that I was safe from the birch, as he invariably went to Durham and spent the day there. But when I saw that he had his flaxen wig on, my countenance fell. He was in the schoolroom all day, and I was too often placed in a very uncomfortable position at nightfall. But sometimes I had to come in contact with the birch-rod for various frolics independent of school erudition. I once smarted severely for an act of kindness. We had a boy named Bryan Salvin, from Croxdale Hall. He was a dull, sluggish, and unwieldy lad, quite incapable of climbing exertions. Being dissatisfied with the regulations of the establishment, he came to me one Palm Sunday, and entreated me to get into the schoolroom through the window, and write a letter of complaint to his sister Eliza in York. I did so, having insinuated myself with vast exertion through the iron stanchions which secured the window; '*sed revocare gradum.*' Whilst I was thrusting might and main through the stanchions, on my way out—suddenly, oh, horrible! the schoolroom door flew open, and on the threshold stood the Reverend Mr. Storey—a fiery, frightful, formidable spectre! To my horror and confusion I drove my foot quite through a pane of glass, and there I stuck, impaled and imprisoned, but luckily not injured by the broken glass. Whilst I was thus in unexpected captivity, he cried out, in an angry voice, 'So you are there, Master Charles, are you?' He got assistance, and they pulled me back by main force. But as this was Palm Sunday my execution was obligingly deferred until Monday morning.

"But let us return to Tudhoe. In my time it was a peaceful, healthy farming village, and abounded in local curiosities. Just on the king's highway, betwixt Durham and Bishop-Auckland, and one field from the school, there stood a public-house called the 'White Horse,' and kept by a man of the name of Charlton. He had a real gaunt English mastiff, half-starved for want of food, and so ferocious that nobody but himself dared to approach it. This publican had also a mare, surprising in her progeny; she had three foals, in three successive years, not one of which had the least appearance of a tail.

"One of Mr. Storey's powdered wigs was of so tempting an aspect, on the shelf where it was laid up in ordinary, that the cat actually kittened in it. I saw her and her little ones all together in the warm wig. He also kept a little white and black bitch, apparently of King Charles's breed. One evening, as we scholars were returning from a walk, Chloe started a hare, which we surrounded and captured, and carried in triumph to oily Mrs. Atkinson, who begged us a play-day for our success.

"On Easter Sunday Mr. Storey always treated us to 'Pasche eggs.' They were boiled hard in a concoction of whin-flowers, which rendered them beautifully purple. We used them for warlike purposes, by holding them betwixt our forefinger and thumb with the sharp end upwards, and as little exposed as possible. An antagonist then approached, and with the sharp end of his own egg struck this egg. If he succeeded in cracking it, the vanquished egg was his; and he either sold it for a halfpenny in the market, or reserved it for his own eating. When all the sharp ends had been crushed, then the blunt ends entered into battle. Thus nearly every Pasche egg in the school had its career of combat. The possessor of a strong egg with a thick shell would sometimes vanquish a dozen of his opponents, all of which the conqueror ultimately transferred into his own stomach, when no more eggs with unbroken ends remained to carry on the war of Easter Week.

"The little black and white bitch once began to snarl, and then to bark at me, when I was on a roving expedition in quest of hens' nests. I took up half a brick and knocked it head over heels. Mr. Storey was watching at the time from one of the upper windows; but I had not seen him, until I heard the sound of his magisterial voice. He beckoned me to his room there and then, and whipped me soundly for my pains.

"Four of us scholars stayed at Tudhoe during the summer vacation, when all the rest had gone home. Two of these had dispositions as malicious as those of two old apes. One fine summer's morning they decoyed me into a field (I was just then from my mother's nursery) where there was a flock of geese. They assured me that the geese had no right to be there; and that it was necessary we should kill them, as they were trespassing on our master's grass. The scamps then furnished me with a hedge-stake. On approaching the flock, behold the gander came out to meet me; and whilst he was hissing defiance at us, I struck him on the neck, and killed him outright. My comrades immediately took to flight, and on reaching the house informed our master of what I had done. But when he heard my unvarnished account of the gander's death, he did not say one single unkind word to me, but scolded most severely the two boys who had led me into the scrape. The geese belonged to a farmer named John Hey, whose son Ralph used to provide me with birds' eggs. Ever after when I passed by his house, some of the children would point to me and say, 'Yaw killed aur guise.'

"At Bishop-Auckland there lived a man by the name of Charles the Painter. He played extremely well on the Northumberland bagpipe, and his neighbour was a good performer on the flageolet. When we had pleased our master by continued good conduct, he would send for these two musicians, who gave us a delightful evening concert in the general play-room, Mr. Storey himself supplying an extra treat of fruit, cakes, and tea.

"Tudhoe had her own ghosts and spectres, just as the neighbouring villages had theirs. One was the Tudhoe mouse, well known and often seen in every house in the village; but I cannot affirm that I myself ever saw it. It was an enormous mouse, of a dark brown colour, and did an immensity of mischief. No cat could face it; and as it wandered through the village, all the dogs would take off, frightened out of their wits, and howling as they ran away. William Wilkinson, Mr. Storey's farming man, told me he had often seen it, but that it terrified him to such a degree that he could not move from the place where he was standing.

"Our master kept a large tom-cat in the house. A fine young man, in the neighbouring village of Ferry-hill, had been severely bitten by a cat, and he died raving mad. On the day that we got this information from Timothy Pickering, the carpenter at Tudhoe, I was on the prowl for adventures, and in passing through Mr. Storey's back kitchen, his big black cat came up to me. Whilst I was tickling its bushy tail, it turned round upon me, and gave me a severe bite in the calf of the leg. This I kept a profound secret, but I was quite sure I should go mad every day, for many months afterwards.

"There was a blacksmith's shop leading down the village to Tudhoe Old Hall. Just opposite this shop was a pond, on the other side of the road. When any sudden death was to take place, or any sudden ill to befall the village, a large black horse used to emerge from it, and walk slowly up and down the village, carrying a rider without a head. The blacksmith's grandfather, his father, himself, his three sons, and two daughters, had seen this midnight apparition rise out of the pond, and return to it before the break of day. John Hickson and Neddy Hunt, two hangers-on at the blacksmith's shop, had seen this phantom more than once, but they never durst approach it. Indeed, every man and woman and child believed in this centaur-spectre, and I am not quite sure if our old master himself did not partly believe that such a thing had occasionally been seen on very dark nights.

"Tudhoe has no river, a misfortune *'valde deflendus.'* In other respects the vicinity was charming; and it afforded an ample supply of woods and hedgerow trees to insure a sufficient stock of carrion crows, jackdaws, jays, magpies, brown owls, kestrels, merlins, and sparrow-hawks, for the benefit of natural history and my own instruction and amusement."

In 1796 Waterton left Tudhoe school and went to Stonyhurst College in Lancashire. It was a country house of the picturesque style of King James I., which had just been made over by Mr. Weld of Lulworth to the Jesuits expelled from Liége. The country round Stonyhurst is varied by hills and streams, and there are mountains at no great distance.

> "Whernside, Pendle Hill, and Ingleboro',
> Three higher hills you'll not find England thoro',"

as they are described, with equal disregard of exact mensuration and of rhythm, in a local rhyme which Waterton learned. Curlew used to fly by in flocks, and the country people had also a rhyme about the curlew:—

> "Be she white or be she black,
> She carries sixpence on her back,"

which Waterton used to say showed how our ancestors valued the bird at table.

At Stonyhurst he read a good deal of Latin and of English literature, and acquired a taste for writing Latin verse. He always looked back on his education there with satisfaction, and in after-life often went to visit the college. Throughout life he never drank wine, and this fortunate habit was the result of the good advice of one of his teachers:—

"My master was Father Clifford, a first cousin of the noble lord of that name. He had left the world, and all its alluring follies, that he might serve Almighty God more perfectly, and work his way with more security up to the regions of eternal bliss. After educating those entrusted to his charge with a care and affection truly paternal, he burst a blood-vessel, and retired to Palermo for the benefit of a warmer climate. There he died the death of the just, in the habit of St. Ignatius.

"One day, when I was in the class of poetry, and which was about two years before I left the college for good and all, he called me up to his room. 'Charles,' said he to me, in a tone of voice perfectly irresistible, 'I have long been studying your disposition, and I clearly foresee that nothing will keep you at home. You will journey into far-distant countries, where you will be exposed to many dangers. There is only one way for you to escape them. Promise me that, from this day forward, you will never put your lips to wine, or to spirituous liquors.' 'The sacrifice is nothing,' added he; 'but, in the end, it will prove of incalculable advantage to you.' I agreed to his enlightened proposal; and from that hour to this, which is now about nine-and-thirty years, I have never swallowed one glass of any kind of wine or of ardent spirits."

After leaving college Waterton stayed at home with his father, and enjoyed fox-hunting for a while. To the end of his days he liked to hear of a good run, and he would now and then look with pleasure on an engraving which hung in the usual dining-room at Walton Hall, representing Lord Darlington, the first master of hounds he had known, well seated on a powerful horse and surrounded by very muscular hounds. In 1802 he went to visit two uncles in Spain, and stayed for more than a year, and there had a terrible experience of pestilence and of earthquake:—

"There began to be reports spread up and down the city that the black vomit had made its appearance; and every succeeding day brought testimony that things were not as they ought to be. I myself, in an alley near my uncles' house, saw a mattress of most suspicious appearance hung out to dry. A Maltese captain, who had dined with us in good health at one o'clock, lay dead in his cabin before sunrise the next morning. A few days after this I was seized with vomiting and fever during the night. I had the most dreadful spasms, and it was supposed that I could not last out till noon the next day. However, strength of constitution got me through it. In three weeks more, multitudes were seen to leave the city, which shortly after was declared to be in a state of pestilence. Some affirmed that the disorder had come from the Levant; others said that it had been imported from the Havanna; but I think it probable that nobody could tell in what quarter it had originated.

"We had now all retired to the country-house—my eldest uncle returning to Malaga from time to time, according as the pressure of business demanded his presence in the city. He left us one Sunday evening, and said he would be back again some time on Monday; but that was my poor uncle's last day's ride. On arriving at his house in Malaga, there was a messenger waiting to inform him that Father Bustamante had fallen sick, and wished to see him. Father Bustamante was an aged priest, who had been particularly kind to my uncle on his first arrival in Malaga. My uncle went immediately to Father Bustamante, gave him every consolation in his power, and then returned to his own house very unwell, there to die a martyr to his charity. Father Bustamante breathed his last before daylight; my uncle took to his bed, and never rose more. As soon as we had received information of his sickness, I immediately set out on foot for the city. His friend, Mr. Power, now of Gibraltar, was already in his room, doing everything that friendship could suggest or prudence dictate. My uncle's athletic constitution bore up against the disease much longer than we thought it possible. He struggled with it for five days, and sank at last about the hour of sunset. He stood six feet four inches high; and was of so kind and generous a disposition, that he was beloved by all who knew him. Many a Spanish tear flowed when it was known that he had ceased to be. We got him a kind of coffin made, in which he was conveyed at midnight to the outskirts of the town, there to be put into one of the pits which the galley-slaves had dug during the day for the reception of the dead. But they could not spare room for the coffin; so the body was taken out of it, and thrown upon the heap which already occupied the pit. A Spanish marquis lay just below him.

"Thousands died as though they had been seized with cholera, others with black vomit, and others of decided yellow fever. There were a few instances of some who departed this life with very little pain or bad symptoms: they felt unwell, they went to bed, they had an idea that they would not get better, and they expired in a kind of slumber. It was sad in the extreme to see the bodies placed in the streets at the close of day, to be ready for the dead-carts as they passed along. The dogs howled fearfully during the night. All was gloom and horror in every street; and you might see the vultures on the strand tugging at the bodies which were washed ashore by the eastern wind. It was always said that 50,000 people left the city at the commencement of the pestilence; and that 14,000 of those who remained in it fell victims to the disease.

"There was an intrigue going on at court, for the interest of certain powerful people, to keep the port of Malaga closed long after the city had been declared free from the disorder; so that none of the vessels in the mole could obtain permission to depart for their destination.

"In the meantime the city was shaken with earthquakes; shock succeeding shock, till we all imagined that a catastrophe awaited us similar to that which had taken place at Lisbon. The pestilence killed you by degrees, and its approaches were sufficiently slow, in general, to enable you to submit to it with firmness and resignation; but the idea of being swallowed up alive by the yawning earth at a moment's notice, made you sick at heart, and rendered you almost fearful of your own shadow. The first shock took place at six in the evening, with a noise as though a thousand carriages had dashed against each other. This terrified many people to such a degree that they paced all night long up and down the Alameda, or public walk, rather than retire to their homes. I went to bed a little after midnight, but was roused by another shock about five o'clock in the morning. It gave the bed a motion which made me fancy that it moved under me from side to side. I sprang up, and having put on my unmentionables (we wore no trousers in those days), I ran out, in all haste, to the Alameda. There the scene was most distressing: multitudes of both sexes, some nearly in a state of nudity, and others sick at stomach, were huddled together, not knowing which way to turn or what to do.

—'Omnes eodem cogimur.'

However, it pleased Heaven, in its mercy, to spare us. The succeeding shocks became weaker and weaker, till at last we felt no more of them."

A courageous sea-captain at last sailed away in safety, though chased by the Spanish brigs of war, and after thirty days at sea Waterton landed in England.

Another uncle had estates in Demerara, and in the autumn Waterton sailed thither from Portsmouth. He landed at Georgetown, Demerara, in

November, 1804, and was soon delighted by the natural history of the tropical forest. In 1806 his father died, and he returned to England. He made four more journeys to Guiana, and, in 1825, published an account of them, entitled "Wanderings in South America, the North-West of the United States, and the Antilles, in the years 1812, 1816, 1820, and 1824; with original instructions for the perfect preservation of birds, &c., for cabinets of natural history." The two first journeys are now reprinted from the original text. The book at once attracted general attention, became popular, and has taken a place among permanent English literature. Unlike most travellers, Waterton tells nothing of his personal difficulties and discomforts, and encumbers his pages with neither statistics nor information of the guidebook kind. His observation of birds and beasts, written down in the forests, and the description of the forests themselves, fill all his pages. The great ant-eater and the sloth were for the first time accurately described by him. He showed that the sloth, instead of being a deformed, unhappy creature, was admirably adapted to its habitat. He explained the use of the great claws of the ant-eater, and the curious gait which they necessitated. The habits of the toucan, of the houtou, of the campanero, and of many other birds, were first correctly described by him. He determined to catch a cayman or alligator, and at last hooked one with a curious wooden hook of four barbs made for him by an Indian.

The adventure which followed is perhaps one of the most famous exploits of an English naturalist.

"We found a cayman, ten feet and a half long, fast to the end of the rope. Nothing now remained to do, but to get him out of the water without injuring his scales, 'hoc opus, hic labor.' We mustered strong: there were three Indians from the creek, there was my own Indian, Yan; Daddy Quashi, [24] the negro from Mrs. Peterson's; James, Mr. R. Edmonstone's man, whom I was instructing to preserve birds; and, lastly, myself.

"I informed the Indians that it was my intention to draw him quietly out of the water, and then secure him. They looked and stared at each other, and said I might do it myself, but they would have no hand in it; the cayman would worry some of us. On saying this, 'consedere duces,' they squatted on their hams with the most perfect indifference.

"The Indians of those wilds have never been subject to the least restraint; and I knew enough of them to be aware, that if I tried to force them against their will, they would take off, and leave me and my presents unheeded, and never return.

"Daddy Quashi was for applying to our guns, as usual, considering them our best and safest friends. I immediately offered to knock him down for his cowardice, and he shrank back, begging that I would be cautious, and not get myself worried; and apologising for his own want of resolution. My Indian was now in conversation with the others, and they asked me if I would allow them to shoot a dozen arrows into him, and thus disable him. This would have ruined all. I had come above three hundred miles on purpose to get a cayman uninjured, and not to carry back a mutilated specimen. I rejected their proposition with firmness, and darted a disdainful eye upon the Indians.

"Daddy Quashi was again beginning to remonstrate, and I chased him on the sand-bank for a quarter of a mile. He told me afterwards, he thought he should have dropped down dead with fright, for he was firmly persuaded, if I had caught him, I should have bundled him into the cayman's jaws. Here then we stood, in silence, like a calm before a thunder-storm. 'Hoc res summa loco. Scinditur in contraria valgus.' They wanted to kill him, and I wanted to take him alive.

"I now walked up and down the sand, revolving a dozen projects in my head. The canoe was at a considerable distance, and I ordered the people to bring it round to the place where we were. The mast was eight feet long, and not much thicker than my wrist. I took it out of the canoe, and wrapped the sail round the end of it. Now it appeared clear to me, that if I went down upon one knee, and held the mast in the same position as the soldier holds his bayonet when rushing to the charge, I could force it down the cayman's throat, should he come open-mouthed at me. When this was told to the Indians, they brightened up, and said they would help me to pull him out of the river.

"'Brave squad!' said I to myself, '"Audax omnia perpeti," now that you have got me betwixt yourselves and danger.' I then mustered all hands for the last time before the battle. We were, four South American savages, two negroes from Africa, a creole from Trinidad, and myself, a white man from Yorkshire. In fact, a little Tower of Babel group, in dress, no dress, address, and language.

"Daddy Quashi hung in the rear; I showed him a large Spanish knife, which I always carried in the waistband of my trousers: it spoke volumes to him, and he shrugged up his shoulders in absolute despair. The sun was just peeping over the high forests on the eastern hills, as if coming to look on, and bid us act with becoming fortitude. I placed all the people at the end of the rope, and ordered them to pull till the cayman appeared on the surface of the water and then, should he plunge, to slacken the rope and let him go again into the deep.

"I now took the mast of the canoe in my hand (the sail being tied round the end of the mast) and sank down upon one knee, about four yards from the water's edge, determining to thrust it down his throat, in case he gave me an opportunity. I certainly felt somewhat uncomfortable in this situation, and I thought of Cerberus on the other side of the Styx ferry. The people pulled the cayman to the surface; he plunged furiously as soon as he arrived in these upper regions, and immediately went below again on their slackening the rope. I saw enough not to fall in love at first sight. I now told them we would run all risks, and have him on land immediately. They pulled again, and out he came—'monstrum horrendum, informe.' This was an interesting moment. I kept my position firmly, with my eye fixed steadfast on him.

"By the time the cayman was within two yards of me, I saw he was in a state of fear and perturbation: I instantly dropped the mast, sprang up, and jumped on his back, turning half round as I vaulted, so that I gained my seat with my face in a right position. I immediately seized his fore-legs, and by main force twisted them on his back; thus they served me for a bridle.

"He now seemed to have recovered from his surprise, and probably fancying himself in hostile company, be began to plunge furiously, and lashed the sand with his long and powerful tail. I was out of reach of the strokes of it, by being near his head. He continued to plunge and strike, and made my seat very uncomfortable. It must have been a fine sight for an unoccupied spectator.

"The people roared out in triumph, and were so vociferous, that it was some time before they heard me tell them to pull me and my beast of burthen farther in. I was apprehensive the rope might break, and then there would have been every chance of going down to the regions under water with the cayman. That would have been more perilous than Arion's marine morning ride:—

> 'Delphini insidens vada cærula sulcat Arion.'

"The people now dragged us about forty yards on the sand; it was the first and last time I was ever on a cayman's back. Should it be asked, how I managed to keep my seat, I would answer—I hunted some years with Lord Darlington's fox-hounds.

"After repeated attempts to regain his liberty, the cayman gave in, and became tranquil through exhaustion. I now managed to do up his jaws, and firmly secured his fore-feet in the position I had held them. We had now another severe struggle for superiority, but he was soon overcome, and again remained quiet. While some of the people were pressing upon his head and shoulders, I threw myself on his tail, and by keeping it down to the sand, prevented him from kicking up another dust. He was finally conveyed to the canoe, and then to the place where we had suspended our hammocks. There I cut his throat; and after breakfast was over, commenced the dissection."

After his fourth journey Waterton occasionally travelled on the Continent, but for the most part resided at Walton Hall. In the park he made the observations afterwards published as "Essays on Natural History," in three series, and since reprinted, with his Life and Letters, by Messrs. Warne and Co.

Walton Hall is situated on an island surrounded by its ancient moat, a lake of about five-and-twenty acres in extent. From the shores of the lake the land rises; parts of the slope, and nearly all the highest part, being covered with wood.

In one wood there was a large heronry, in another a rookery. Several hollow trees were haunted by owls, in the summer goat-suckers were always to be seen in the evening flying about two oaks on the hill. At one end of the lake in summer the kingfisher might be watched fishing, and throughout the year herons waded round its shores picking up fresh-water mussels, or stood motionless for hours, watching for fish. In winter, when the lake was frozen, three or four hundred wild duck, with teal and pochards, rested on it all day, and flew away at night to feed; while widgeons fed by day on its shores. Coots and water-hens used to come close to the windows and pick up food put out for them. The Squire built a wall nine feet high all round his park, and he used laughingly to say that he paid for it with the cost of the wine which he did not drink after dinner.

A more delightful home for a naturalist could not have been. No shot was ever fired within the park wall, and every year more birds came. Waterton used often to quote the lines:—

> "No bird that haunts my valley free
>   To slaughter I condemn;
> Taught by the Power that pities me,
>   I learn to pity them;"

and each new-comer added to his happiness. In his latter days the household usually consisted of the Squire, as he was always called, and of his two sisters-in-law, for he had lost his wife soon after his marriage in 1829. He breakfasted at eight, dined in the middle of the day, and drank tea in the evening. He went to bed early, and slept upon the bare floor, with a block of wood for his pillow. He rose for the day at half-past three, and spent the hour from four to five at prayer in his chapel. He then read every morning a chapter in a Spanish Life of St. Francis Xavier, followed by a chapter of "Don Quixote" in the original, after which he used to stuff birds or write letters till breakfast. Most of the day he spent in the open air, and when the weather was cold would light a fire of sticks and warm himself by it. So active did he continue to the end of his days, that on his eightieth birthday he climbed an oak in my company. He was very kind to the poor, and threw open a beautiful part of his park to excursionists all through the summer. He had a very tender heart for beasts and birds, as well as for men. If a cat looked hungry he would see that she had a meal, and sometimes when he had forgotten to put a crust of bread in his pocket before starting on his afternoon walk, he would say to his companion, "How shall we ever get past that goose?" for there was a goose which used to wait for him in the evening at the end of the bridge over the moat, and he could not bear to disappoint it. If he could not find a bit of food for it, he would wait at a distance till the bird went away, rather than give it nothing when it raised its bill.

Towards the end of his life I enjoyed his friendship, and can never forget his kindly welcome, his pithy conversation, the happy humour with which he expressed the conclusions of his long experience of men, birds and beasts, and the goodness which shone from his face. I was staying at Walton when he died, and have thus described his last hours in the biography which is prefixed to the latest edition of his Essays. [31] I was reading for an examination, and used, on the Squire's invitation, to go and chat with him just after midnight, for at that hour be always awoke, and paid a short visit to his chapel. A little before midnight on May 24th I visited him in his room. He was sitting asleep by his fire wrapped up in a large Italian cloak.

His head rested upon his wooden pillow, which was placed on a table, and his thick silvery hair formed a beautiful contrast with the dark colour of the oak. He soon woke up, and withdrew to the chapel, and on his return we talked together for three-quarters of an hour about the brown owl, the nightjar, and other birds. The next morning, May 25, he was unusually cheerful, and said to me, "That was a very pleasant little confab we had last night: I do not suppose there was such another going on in England at the same time." After breakfast we went with a carpenter to finish some bridges at the far end of the park. The work was completed, and we were proceeding homewards when, in crossing a small bridge, a bramble caught the Squire's foot, and he fell heavily upon a log. He was greatly shaken, and said he thought he was dying. He walked, notwithstanding, a little way, and was then compelled to lie down. He would not permit his sufferings to distract his mind, and he pointed out to the carpenter some trees which were to be felled. He presently continued his route, and managed to reach the spot where the boat was moored. Hitherto he had refused all assistance, but he could not step from the bank into the boat, and he said, "I am afraid I must ask you to help me in." He walked from the landing-place into the house, changed his clothes, and came and sat in the large room below. The pain increasing, he rose from his seat after he had seen his doctor, and though he had been bent double with anguish, he persisted in walking up-stairs without help, and would have gone to his own room in the top storey, if, for the sake of saving trouble to others, he had not been induced to stop half-way in Miss Edmonstone's sitting-room. Here he lay down upon the sofa, and was attended by his sisters-in-law. The pain abated, and the next day he seemed better. In the afternoon he talked to me a good deal, chiefly about natural history. But he was well aware of his perilous condition, for he remarked to me, "This is a bad business," and later on he felt his pulse often, and said, "It is a bad case." He was more than self-possessed. A benignant cheerfulness beamed from his mind, and in the fits of pain he frequently looked up with a gentle smile, and made some little joke. Towards midnight he grew worse. The priest, the Reverend R. Browne, was summoned, and Waterton got ready to die. He pulled himself upright without help, sat in the middle of the sofa, and gave his blessing in turn to his grandson, Charlie, to his granddaughter, Mary, to each of his sisters-in-law, to his niece, and to myself, and left a message for his son, who was hastening back from Rome. He then received the last sacraments, repeated all the responses, Saint Bernard's hymn in English, and the first two verses of the *Dies Iræ*. The end was now at hand, and he died at twenty-seven minutes past two in the morning of May 27, 1865. The window was open. The sky was beginning to grow grey, a few rooks had cawed, the swallows were twittering, the landrail was craking from the Ox-close, and a favourite cock, which he used to call his morning gun, leaped out from some hollies, and gave his accustomed crow. The ear of his master was deaf to the call. He had obeyed a sublimer summons, and had woke up to the glories of the eternal world.

He was buried on his birthday, the 3rd of June, between two great oaks at the far end of the lake, the oldest trees in the park. He had put up a rough stone cross to mark the spot where he wished to be buried. Often on summer days he had sat in the shade of these oaks watching the kingfishers. "Cock Robin and the magpies," he said to me as we sat by the trees one day, "will mourn my loss, and you will sometimes remember me when I lie here." At the foot of the cross is a Latin inscription which he wrote himself. It could hardly be simpler: "Pray for the soul of Charles Waterton, whose tired bones are buried near this cross." The dates of his birth and death are added.

Walton Hall is no longer the home of the Watertons, the oaks are too old to flourish many years more, and in time the stone cross may be overthrown and the exact burial place of Waterton be forgotten; but his "Wanderings in South America" and his "Natural History Essays" will always be read, and are for him a memorial like that claimed by the poet he read oftenest—

> "quod nec Jovis ira, nec ignes,
> Nec poterit ferrum, nec edax abolere vetustas."

NORMAN MOORE.

# First Journey.

>—"nec herba, nec latens in asperis
>Radix fefellit me locis."

In the month of April, 1812, I left the town of Stabroek, to travel through the wilds of Demerara and Essequibo, a part of *ci-devant* Dutch Guiana, in South America.

The chief objects in view were to collect a quantity of the strongest wourali-poison; and to reach the inland frontier fort of Portuguese Guiana.

It would be a tedious journey for him who wishes to travel through these wilds to set out from Stabroek on foot. The sun would exhaust him in his attempts to wade through the swamps, and the mosquitos at night would deprive him of every hour of sleep.

The road for horses runs parallel to the river, but it extends a very little way, and even ends before the cultivation of the plantation ceases.

The only mode then that remains is to proceed by water; and when you come to the high lands, you may make your way through the forest on foot, or continue your route on the river.

After passing the third island in the river Demerara, there are few plantations to be seen, and those not joining on to one another, but separated by large tracts of wood.

The Loo is the last where the sugar-cane is growing. The greater part of its negroes have just been ordered to another estate; and ere a few months shall have elapsed all signs of cultivation will be lost in under-wood.

Higher up stand the sugar-works of Amelia's Waard, solitary and abandoned; and after passing these there is not a ruin to inform the traveller that either coffee or sugar has been cultivated.

From Amelia's Waard an unbroken range of forest covers each bank of the river, saving here and there where a hut discovers itself, inhabited by free people of colour, with a rood or two of bared ground about it; or where the wood-cutter has erected himself a dwelling, and cleared a few acres for pasturage. Sometimes you see level ground on each side of you for two or three hours at a stretch; at other times a gently sloping hill presents itself; and often, on turning a point, the eye is pleased with the contrast of an almost perpendicular height jutting into the water. The trees put you in mind of an eternal spring, with summer and autumn kindly blended into it.

Here you may see a sloping extent of noble trees, whose foliage displays a charming variety of every shade from the lightest to the darkest green and purple. The tops of some are crowned with bloom of the loveliest hue; while the boughs of others bend with a profusion of seeds and fruits.

Those whose heads have been bared by time, or blasted by the thunderstorm, strike the eye, as a mournful sound does the ear in music; and seem to beckon to the sentimental traveller to stop a moment or two, and see that the forests which surround him, like men and kingdoms, have their periods of misfortune and decay.

The first rocks of any considerable size that are observed on the side of the river are at a place called Saba, from the Indian word, which means a stone. They appear sloping down to the water's edge, not shelvy, but smooth, and their exuberances rounded off, and, in some places, deeply furrowed, as though they had been worn with continual floods of water.

There are patches of soil up and down, and the huge stones amongst them produce a pleasing and novel effect. You see a few coffee-trees of a fine luxuriant growth; and nearly on the top of Saba stands the house of the post-holder.

He is appointed by government to give in his report to the protector of the Indians of what is going on amongst them, and to prevent suspicious people from passing up the river.

When the Indians assemble here the stranger may have an opportunity of seeing the aborigines dancing to the sound of their country music, and painted in their native style. They will shoot their arrows for him with an unerring aim, and send the poisoned dart from the blow-pipe true to its destination; and here he may often view all the different shades, from the red savage to the white man, and from the white man to the sootiest son of Africa.

Beyond this post there are no more habitations of white men, or free people of colour.

In a country so extensively covered with wood as this is, having every advantage that a tropical sun and the richest mould, in many places, can give to vegetation, it is natural to look for trees of very large dimensions; but it is rare to meet with them above six yards in circumference. If larger have ever existed, they had fallen a sacrifice either to the axe or to fire.

If, however, they disappoint you in size, they make ample amends in height. Heedless and bankrupt in all curiosity must he be who can journey on without stopping to take a view of the towering mora. Its topmost branch, when naked with age or dried by accident, is the favourite resort of the toucan. Many a time has this singular bird felt the shot faintly strike him from the gun of the fowler beneath, and owed his life to the distance betwixt them.

The trees which form these far-extending wilds are as useful as they are ornamental. It would take a volume of itself to describe them.

The green-heart, famous for its hardness and durability; the hackea, for its toughness; the ducalabali, surpassing mahogany; the ebony and letter-wood vieing with the choicest woods of the old world; the locust tree, yielding copal; and the hayawa and olon trees, furnishing a sweet-smelling resin, are all to be met with in the forest, betwixt the plantations and the rock Saba.

Beyond this rock the country has been little explored; but it is very probable that these, and a vast collection of other kinds, and possibly many new species, are scattered up and down, in all directions, through the swamps, and hills, and savannas of *ci-devant* Dutch Guiana.

On viewing the stately trees around him the naturalist will observe many of them bearing leaves, and blossoms, and fruit, not their own.

The wild fig-tree, as large as a common English apple-tree, often rears itself from one of the thick branches at the top of the mora; and when its fruit is ripe, to it the birds resort for nourishment. It was to an undigested seed, passing through the body of the bird which had perched on the mora, that the fig-tree first owed its elevated station there. The sap of the mora raised it into full bearing; but now, in its turn, it is doomed to contribute a portion of its own sap and juices towards the growth of different species of vines, the seeds of which, also, the birds deposited on its branches. These soon vegetate, and bear fruit in great quantities; so what with their usurpation of the resources of the fig-tree, and the fig-tree of the mora, the mora, unable to support a charge which nature never intended it should, languishes and dies under its burden; and then the fig-tree, and its usurping progeny of vines, receiving no more succour from their late foster-parent, droop and perish in their turn.

A vine called the bush-rope by the wood-cutters, on account of its use in hauling out the heaviest timber, has a singular appearance in the forests of Demerara. Sometimes you see it nearly as thick as a man's body, twisted like a cork-screw round the tallest trees, and rearing its head high above their tops. At other times three or four of them, like strands in a cable, join tree and tree and branch and branch together. Others, descending from on high, take root as soon as their extremity touches the ground, and appear like shrouds and stays supporting the main-mast of a line-of-battle ship; while others, sending out parallel, oblique, horizontal, and perpendicular shoots in all directions, put you in mind of what travellers call a matted forest. Oftentimes a tree, above a hundred feet high, uprooted by the whirlwind, is stopped in its fall by these amazing cables of nature; and hence it is that you account for the phenomenon of seeing trees not only vegetating, but sending forth vigorous shoots, though far from their perpendicular, and their trunks inclined to every degree from the meridian to the horizon.

Their heads remain firmly supported by the bush-rope; many of their roots soon refix themselves in the earth, and frequently a strong shoot will sprout out perpendicularly from near the root of the reclined trunk, and in time become a fine tree. No grass grows under the trees; and few weeds, except in the swamps.

The high grounds are pretty clear of underwood, and with a cutlass to sever the small bush-ropes, it is not difficult walking among the trees.

The soil, chiefly formed by the fallen leaves and decayed trees, is very rich and fertile in the valleys. On the hills it is little better than sand. The rains seem to have carried away and swept into the valleys every particle which nature intended to have formed a mould.

Four-footed animals are scarce, considering how very thinly these forests are inhabited by men.

Several species of the animal commonly called tiger, though in reality it approaches nearer to the leopard, are found here; and two of their diminutives, named tiger-cats. The tapir, the labba, and deer, afford excellent food, and chiefly frequent the swamps and low ground, near the sides of the river and creeks.

In stating that four-footed animals are scarce, the peccary must be excepted. Three or four hundred of them herd together, and traverse the wilds in all directions, in quest of roots and fallen seeds. The Indians mostly shoot them with poisoned arrows. When wounded, they run about one hundred and fifty paces; they then drop, and make wholesome food.

The red monkey, erroneously called the baboon, is heard oftener than it is seen; while the common brown monkey, the bisa, and sacawinki, rove from tree to tree, and amuse the stranger as he journeys on.

A species of the polecat, and another of the fox, are destructive to the Indian's poultry; while the opossum, the guana, and salempenta afford him a delicious morsel.

The small ant-bear, and the large one, remarkable for its long, broad bushy tail, are sometimes seen on the tops of the wood-ants' nests; the armadillos bore in the sand-hills, like rabbits in a warren; and the porcupine is now and then discovered in the trees over your head.

This, too, is the native country of the sloth. His looks, his gestures, and his cries, all conspire to entreat you to take pity on him. These are the only weapons of defence which nature has given him. While other animals assemble in herds, or in pairs range through these boundless wilds, the sloth is solitary, and almost stationary; he cannot escape from you. It is said, his piteous moans make the tiger relent, and turn out of the way. Do not then level your gun at him, or pierce him with a poisoned arrow; he has never hurt one living creature. A few leaves, and those of the commonest and coarsest kind, are all he asks for his support. On comparing him with other animals, you would say that you could perceive deficiency, deformity, and superabundance in his composition. He has no cutting teeth, and though four stomachs, he still wants the long intestines of ruminating animals. He has only one inferior aperture, as in birds. He has no soles to his feet, nor has he the power of moving his toes separately. His hair is flat, and puts you in mind of grass withered by the wintry blast. His legs are too short; they appear deformed by the manner in which they are joined to the body; and when he is on the ground, they seem as if only calculated to be of use in climbing trees. He has forty-six ribs, while the elephant has only forty; and his claws are disproportionably long. Were you to mark down, upon a graduated scale, the different claims to superiority amongst the four-footed animals, this poor ill-formed creature's claim would be the last upon the lowest degree.

Demerara yields to no country in the world in her wonderful and beautiful productions of the feathered race. Here the finest precious stones are far surpassed by the vivid tints which adorn the birds. The naturalist may exclaim, that nature has not known where to stop in forming new species, and painting her requisite shades. Almost every one of those singular and elegant birds described by Buffon as belonging to Cayenne are to be met with in Demerara; but it is only by an indefatigable naturalist that they are to be found.

The scarlet curlew breeds in innumerable quantities in the muddy islands on the coasts of Pomauron; the egrets and crabiers in the same place. They resort to the mud-flats at ebbing water, while thousands of sandpipers and plovers, with here and there a spoonbill and flamingo, are seen amongst them. The pelicans go farther out to sea, but return at sundown to the courada trees. The humming-birds are chiefly to be found near the flowers at which each of the species of the genus is wont to feed. The pie, the gallinaceous, the columbine, and passerine tribes, resort to the fruit-bearing trees.

You never fail to see the common vulture where there is carrion. In passing up the river there was an opportunity of seeing a pair of the king of the vultures; they were sitting on the naked branch of a tree, with about a dozen of the common ones with them. A tiger had killed a goat the day before; he had been driven away in the act of sucking the blood, and not finding it safe or prudent to return, the goat remained in the same place where he had killed it; it had begun to putrefy, and the vultures had arrived that morning to claim the savoury morsel.

At the close of day, the vampires leave the hollow trees, whither they had fled at the morning's dawn, and scour along the river's banks in quest of prey. On waking from sleep, the astonished traveller finds his hammock all stained with blood. It is the vampire that has sucked him. Not man alone, but every unprotected animal, is exposed to his depredations; and so gently does this nocturnal surgeon draw the blood, that instead of being roused, the patient is lulled into a still profounder sleep. There are two species of vampire in Demerara, and both suck living animals; one is rather larger than the common bat; the other measures above two feet from wing to wing extended.

Snakes are frequently met with in the woods betwixt the sea-coast and the rock Saba, chiefly near the creeks and on the banks of the river. They are large, beautiful, and formidable. The rattlesnake seems partial to a tract of ground known by the name of Canal Number Three; there the effects of his poison will be long remembered.

The camoudi snake has been killed from thirty to forty feet long; though not venomous, his size renders him destructive to the passing animals. The Spaniards in the Oroonoque positively affirm that he grows to the length of seventy or eighty feet, and that he will destroy the strongest and largest bull. His name seems to confirm this; there he is called "matatoro," which literally means "bull-killer." Thus he may be ranked amongst the deadly snakes: for it comes nearly to the same thing in the end, whether the victim dies by poison from the fangs, which corrupts his blood and makes it stink horribly, or whether his body be crushed to mummy and swallowed by this hideous beast.

The whipsnake, of a beautiful changing green, and the coral with alternate broad transverse bars of black and red, glide from bush to bush, and may be handled with safety; they are harmless little creatures.

The labarri snake is speckled, of a dirty brown colour, and can scarcely be distinguished from the ground or stump on which he is coiled up; he grows to the length of about eight feet, and his bite often proves fatal in a few minutes.

Unrivalled in his display of every lovely colour of the rainbow, and unmatched in the effects of his deadly poison, the counacouchi glides undaunted on, sole monarch of these forests; he is commonly known by the name of the bush-master. Both man and beast fly before him, and allow him to pursue an undisputed path. He sometimes grows to the length of fourteen feet.

A few small caymans, from two to twelve feet long, may be observed now and then in passing up and down the river: they just keep their heads above the water, and a stranger would not know them from a rotten stump.

Lizards of the finest green, brown, and copper colour, from two inches to two feet and a half long, are ever and anon rustling among the fallen leaves, and crossing the path before you; whilst the chameleon is busily employed in chasing insects round the trunks of the neighbouring trees.

The fish are of many different sorts, and well-tasted, but not, generally speaking, very plentiful. It is probable that their numbers are considerably thinned by the otters, which are much larger than those of Europe. In going through the overflowed savannas which have all a communication with the river, you may often see a dozen or two of them sporting among the sedges before you.

This warm and humid climate seems particularly adapted to the producing of insects; it gives birth to myriads, beautiful past description in their variety of tints, astonishing in their form and size, and many of them noxious in their qualities.

He whose eye can distinguish the various beauties of uncultivated nature, and whose ear is not shut to the wild sounds in the woods, will be delighted in passing up the river Demerara. Every now and then, the maam or tinamou sends forth one long and plaintive whistle from the depths of the forest, and then stops; whilst the yelping of the toucan, and the shrill voice of the bird called pi-pi-yo, is heard during the interval. The campanero never fails to attract the attention of the passenger: at a distance of nearly three miles you may hear this snow-white bird tolling every four or five minutes, like the distant convent bell. From six to nine in the morning the forests resound with the mingled cries and strains of the feathered race; after this they gradually die away. From eleven to three all nature is hushed as in a midnight silence, and scarce a note is heard, saving that of the campanero and the pi-pi-yo; it is then that, oppressed by the solar heat, the birds retire to the thickest shade and wait for the refreshing cool of evening.

At sun-down the vampires, bats, and goat-suckers dart from their lonely retreat, and skim along the trees on the river's bank. The different kinds of frogs almost stun the ear with their hoarse and hollow-sounding croaking, while the owls and goat-suckers lament and mourn all night long.

About two hours before daybreak you will hear the red monkey moaning as though in deep distress; the houtou, a solitary bird, and only found in the thickest recesses of the forest, distinctly articulates, "houtou, houtou," in a low and plaintive tone, an hour before sunrise; the maam whistles about the same hour; the hannaquoi, pataca, and maroudi announce his near approach to the eastern horizon, and the parrots and parroquets confirm his arrival there.

The crickets chirp from sunset to sunrise, and often during the day, when the weather is cloudy. The bêterouge is exceeding numerous in these extensive wilds, and not only man, but beasts and birds, are tormented by it. Mosquitos are very rare after you pass the third island in the Demerara, and sand-flies but seldom appear.

Courteous reader, here thou hast the outlines of an amazing landscape given thee; thou wilt see that the principal parts of it are but faintly traced, some of them scarcely visible at all, and that the shades are wholly wanting. If thy soul partakes of the ardent flame which the persevering Mungo Park's did, these outlines will be enough for thee: for they will give some idea of what a noble country this is: and if thou hast but courage to set about giving the world a finished picture of it, neither materials to work on, nor colours to paint it in its true shades, will be wanting to thee. It may appear a difficult task at a distance; but look close at it, and it is nothing at all; provided thou hast but a quiet mind, little more is necessary, and the Genius which presides over these wilds will kindly help thee through the rest. She will allow thee to slay the fawn, and to cut down the mountain-cabbage for thy support, and to select from every part of her domain whatever may be necessary for the work thou art about; but having killed a pair of doves in order to enable thee to give mankind a true and proper description of them, thou must not destroy a third through wantonness, or to show what a good marksman thou art; that would only blot the picture thou art finishing, not colour it.

Though retired from the haunts of men, and even without a friend with thee, thou wouldst not find it solitary. The crowing of the hannaquoi will sound in thine ears like the daybreak town-clock; and the wren and the thrush will join with thee in thy matin hymn to thy Creator, to thank Him for thy night's rest.

At noon the Genius will lead thee to the troely, one leaf of which will defend thee from both sun and rain. And if, in the cool of the evening, thou hast been tempted to stray too far from thy place of abode, and art deprived of light to write down the information thou hast collected, the firefly, which thou wilt see in almost every bush around thee, will be thy candle. Hold it over thy pocket-book, in any position which thou knowest will not hurt it, and it will afford thee ample light. And when thou hast done with it, put it kindly back again on the next branch to thee. It will want no other reward for its services.

When in thy hammock, should the thought of thy little crosses and disappointments, in thy ups and downs through life, break in upon thee, and throw thee into a pensive mood, the owl will bear thee company. She will tell thee that hard has been her fate too; and, at intervals, "Whip-poor-Will" and "Willy come go" will take up the tale of sorrow. Ovid has told thee how the owl once boasted the human form, and lost it for a very small offence; and were the poet alive now, he would inform thee that "Whip-poor-Will," and "Willy come go," are the shades of those poor African and Indian slaves, who died worn out and broken-hearted. They wail and cry, "Whip-poor-Will," "Willy come go," all night long; and often when the moon shines you see them sitting on the green turf, near the houses of those whose ancestors tore them from the bosom of their helpless families, which all probably perished through grief and want after their support was gone.

About an hour above the rock of Saba stands the habitation of an Indian, called Simon, on the top of a hill. The side next the river is almost perpendicular, and you may easily throw a stone over to the opposite bank. Here there was an opportunity of seeing man in his rudest state. The Indians who frequented this habitation, though living in the midst of woods, bore evident marks of attention to their persons. Their hair was neatly collected, and tied up in a knot; their bodies fancifully painted red, and the paint was scented with hayawa. This gave them a gay and animated appearance. Some of them had on necklaces, composed of the teeth of wild boars slain in the chase; many wore rings, and others had an ornament on the left arm, midway betwixt the shoulder and the elbow. At the close of day they regularly bathed in the river below; and the next morning seemed busy in renewing the faded colours of their faces.

One day there came into the hut a form which literally might be called the wild man of the woods. On entering, he laid down a ball of wax, which he had collected in the forest. His hammock was all ragged and torn; and his bow, though of good wood, was without any ornament or polish; "erubuit domino, cultior esse suo." His face was meagre, his looks forbidding, and his whole appearance neglected. His long black hair hung from his head in matted confusion; nor had his body to all appearance ever been painted. They gave him some cassava bread and boiled fish, which he ate voraciously, and soon after left the hut. As he went out you could observe no traces in his countenance or demeanour which indicated that he was in the least mindful of having been benefited by the society he was just leaving.

The Indians said that he had neither wife, nor child, nor friend. They had often tried to persuade him to come and live amongst them; but all was of no avail. He went roving on, plundering the wild bees of their honey, and picking up the fallen nuts and fruits of the forest. When he fell in with game, he procured fire from two sticks, and cooked it on the spot. When a hut happened to be in his way, he stepped in and asked for something to eat, and then months elapsed ere they saw him again. They did not know what had caused him to be thus unsettled; he had been so for years; nor did they believe that even old age itself would change the habits of this poor, harmless, solitary wanderer.

From Simon's, the traveller may reach the large fall with ease in four days.

The first falls that he meets are merely rapids, scarce a stone appearing above the water in the rainy season; and those in the bed of the river barely high enough to arrest the water's course, and by causing a bubbling, show that they are there.

With this small change of appearance in the stream, the stranger observes nothing new till he comes within eight or ten miles of the great fall. Each side of the river presents an uninterrupted range of wood, just as it did below. All the productions found betwixt the plantations and the rock Saba are to be met with here.

From Simon's to the great fall there are five habitations of the Indians—two of them close to the river's side; the other three a little way in the forest. These habitations consist of from four to eight huts, situated on about an acre of ground which they have cleared from the surrounding woods. A few pappaw, cotton, and mountain cabbage-trees are scattered round them.

At one of these habitations a small quantity of the wourali-poison was procured. It was in a little gourd. The Indian who had it said that he had killed a number of wild hogs with it and two tapirs. Appearances seemed to confirm what he had said; for on one side it had been nearly taken out to the bottom at different times, which probably would not have been the case had the first or second trial failed.

Its strength was proved on a middle-sized dog. He was wounded in the thigh, in order that there might be no possibility of touching a vital part. In three or four minutes he began to be affected, smelt at every little thing on the ground around him, and looked wistfully at the wounded part. Soon after this he staggered, laid himself down, and never rose more. He barked once, though not as if in pain. His voice was low and weak; and in a second attempt it quite failed him. He now put his head betwixt his forelegs, and raising it slowly again, he fell over on his side. His eyes immediately became fixed, and though his extremities every now and then shot convulsively, he never showed the least desire to raise up his head. His heart fluttered much from the time he lay down, and at intervals beat very strong; then stopped for a moment or two, and then beat again; and continued faintly beating several minutes after every other part of his body seemed dead.

In a quarter of an hour after he had received the poison he was quite motionless.

A few miles before you reach the great fall, and which, indeed, is the only one which can be called a fall, large balls of froth come floating past you. The river appears beautifully marked with streaks of foam, and on your nearer approach the stream is whitened all over.

At first, you behold the fall rushing down a bed of rocks, with a tremendous noise, divided into two foamy streams, which at their junction again form a small island covered with wood. Above this island, for a short space, there appears but one stream all white with froth, and fretting and boiling amongst the huge rocks which obstruct its course.

Higher up it is seen dividing itself into a short channel or two, and trees grow on the rocks which caused its separation. The torrent in many places has eaten deep into the rocks, and split them into large fragments by driving others against them. The trees on the rocks are in bloom and vigour, though their roots are half bared, and many of them bruised and broken by the rushing waters.

This is the general appearance of the fall from the level of the water below to where the river is smooth and quiet above. It must be remembered that this is during the periodical rains. Probably in the dry season it puts on a very different appearance. There is no perpendicular fall of water of any consequence throughout it, but the dreadful roaring and rushing of the torrent down a long, rocky, and moderately sloping channel has a fine effect; and the stranger returns well pleased with what he has seen. No animal, nor craft of any kind, could stem this downward flood. In a few moments the first would be killed, the second dashed in pieces.

The Indians have a path alongside of it, through the forest, where prodigious crabwood trees grow. Up this path they drag their canoes, and launch them into the river above; and on their return bring them down the same way.

About two hours below this fall is the habitation of an Acoway chief called Sinkerman. At night you hear the roaring of the fall from it. It is pleasantly situated on the top of a sand-hill. At this place you have the finest view the river Demerara affords: three tiers of hills rise in slow gradation, one above the other before you, and present a grand and magnificent scene, especially to him who has been accustomed to a level country.

Here, a little after midnight on the first of May, was heard a most strange and unaccountable noise; it seemed as though several regiments were engaged, and musketry firing with great rapidity. The Indians, terrified beyond description, left their hammocks and crowded all together, like sheep at the approach of the wolf. There were no soldiers within three or four hundred miles. Conjecture was of no avail, and all conversation next morning on the subject was as useless and unsatisfactory as the dead silence which succeeded to the noise.

He who wishes to reach the Macoushi country had better send his canoe over land from Sinkerman's to the Essequibo.

There is a pretty good path, and meeting a creek about three-quarters of the way, it eases the labour, and twelve Indians will arrive with it in the Essequibo in four days.

The traveller need not attend his canoe; there is a shorter and a better way. Half an hour below Sinkerman's he finds a little creek on the western bank of the Demerara. After proceeding about a couple of hundred yards up it, he leaves it, and pursues a west-north-west direction by land for the Essequibo. The path is good, though somewhat rugged with the roots of trees, and here and there obstructed by fallen ones; it extends more over level ground than otherwise. There are a few steep ascents and descents in it, with a little brook running at the bottom of them; but they are easily passed over, and the fallen trees serve for a bridge.

You may reach the Essequibo with ease in a day and a half; and so matted and interwoven are the tops of the trees above you, that the sun is not felt once all the way, saving where the space which a newly-fallen tree occupied lets in his rays upon you. The forest contains an abundance of wild hogs, lobbas, acouries, powisses, maams, maroudis, and waracabas for your nourishment, and there are plenty of leaves to cover a shed whenever you are inclined to sleep.

The soil has three-fourths of sand in it, till you come within half an hour's walk of the Essequibo, where you find a red gravel and rocks. In this retired and solitary tract, nature's garb, to all appearance, has not been injured by fire, nor her productions broken in upon by the exterminating hand of man.

Here the finest green-heart grows, and wallaba, purple-heart, siloabali, sawari, buletre, tauronira, and mora, are met with in vast abundance, far and near, towering up in majestic grandeur, straight as pillars sixty or seventy feet high, without a knot, or branch.

Traveller, forget for a little while the idea thou hast of wandering farther on, and stop and look at this grand picture of vegetable nature; it is a reflection of the crowd thou hast lately been in, and though a silent monitor, it is not a less eloquent one on that account. See that noble purple-heart before thee! Nature has been kind to it. Not a hole, not the least oozing from its trunk, to show that its best days are past. Vigorous in youthful blooming beauty, it stands the ornament of these sequestered wilds, and tacitly rebukes those base ones of thine own species who have been hardy enough to deny the existence of Him who ordered it to flourish there.

Behold that one next to it!—Hark! how the hammerings of the red-headed woodpecker resound through its distempered boughs! See what a quantity of holes he has made in it, and how its bark is stained with the drops which trickle down from them. The lightning, too, has blasted one side of it. Nature looks pale and wan in its leaves, and her resources are nearly dried up in its extremities; its sap is tainted; a mortal sickness, slow as a consumption, and as sure in its consequences, has long since entered its frame, vitiating and destroying the wholesome juices there.

Step a few paces aside, and cast thine eye on that remnant of a mora behind it. Best part of its branches, once so high and ornamental, now lie on the ground in sad confusion one upon the other, all shattered and fungus-grown, and a prey to millions of insects, which are busily employed in destroying them. One branch of it still looks healthy! Will it recover? No, it cannot; nature has already run her course, and that healthy looking branch is only as a fallacious good symptom in him who is just about to die of a mortification when he feels no more pain, and fancies his distemper has left him; it is as the momentary gleam of a wintry sun's ray close to the western horizon.—See! while we are speaking, a gust of wind has brought the tree to the ground, and made room for its successor.

Come farther on, and examine that apparently luxuriant tauronira on thy right hand. It boasts a verdure not its own; they are false ornaments it wears; the bush-rope and bird-vines have clothed it from the root to its topmost branch. The succession of fruit which it hath borne, like good cheer in the houses of the great, has invited the birds to resort to it, and they have disseminated beautiful, though destructive, plants on its branches, which, like the distempers vice brings into the human frame, rob it of all its health and vigour; they have shortened its days, and probably in another year they will finally kill it, long before nature intended that it should die.

Ere thou leavest this interesting scene, look on the ground around thee, and see what everything here below must come to.

Behold that newly fallen wallaba! The whirlwind has uprooted it in its prime, and it has brought down to the ground a dozen small ones in its fall. Its bark has already begun to drop off! And that heart of mora close by it is fast yielding, in spite of its firm, tough texture.

The tree which thou passedst but a little ago, and which perhaps has lain over yonder brook for years, can now hardly support itself, and in a few months more it will have fallen into the water.

Put thy foot on that large trunk thou seest to the left. It seems entire amid the surrounding fragments. Mere outward appearance, delusive phantom of what it once was! Tread on it, and like the fuss-ball, it will break into dust.

Sad and silent mementoes to the giddy traveller as he wanders on! Prostrate remnants of vegetable nature, how incontestably ye prove what we must all at last come to, and how plain your mouldering ruins show that the firmest texture avails us nought when Heaven wills that we should cease to be!—

> "The cloud-capp'd towers, the gorgeous palaces,
> The solemn temples, the great globe itself,
> Yea, all which it inherit, shall dissolve,
> And, like the baseless fabric of a vision,
> Leave not a rack behind."

Cast thine eye around thee, and see the thousands of nature's productions. Take a view of them from the opening seed on the surface, sending a downward shoot, to the loftiest and the largest trees, rising up and blooming in wild luxuriance; some side by side, others separate; some curved and knotty, others straight as lances, all in beautiful gradation, fulfilling the mandates they had received from Heaven, and though condemned to die, still never failing to keep up their species till time shall be no more.

Reader, must thou not be induced to dedicate a few months to the good of the public, and examine with thy scientific eye the productions which the vast and well-stored colony of Demerara presents to thee?

What an immense range of forest is there from the rock Saba to the great fall! and what an uninterrupted extent before thee from it to the banks of the Essequibo! No doubt, there is many a balsam and many a medicinal root yet to be discovered, and many a resin, gum, and oil yet unnoticed. Thy work would be a pleasing one, and thou mightest make several useful observations in it.

Would it be thought impertinent in thee to hazard a conjecture, that, with the resources the Government of Demerara has, stones might be conveyed from the rock Saba to Stabroek to stem the equinoctial tides, which are for ever sweeping away the expensive wooden piles around the mounds of the fort? Or would the timber-merchant point at thee in passing by, and call thee a descendant of La Mancha's knight, because thou maintainest that the stones which form the rapids might be removed with little expense, and thus open the navigation to the woodcutter from Stabroek to the great fall? Or wouldst thou be deemed enthusiastic or biassed, because thou givest it as thy opinion that the climate in these high lands is exceedingly wholesome, and the lands themselves capable of nourishing and maintaining any number of settlers? In thy dissertation on the Indians, thou mightest hint, that possibly they could be induced to help the new settlers a little; and that, finding their labours well requited, it would be the means of their keeping up a constant communication with us, which probably might he the means of laying the first stone towards their Christianity. They are a poor, harmless, inoffensive set of people, and their wanderings and ill-provided way of living seem more to ask for pity from us, than to fill our heads with thoughts that they would be hostile to us.

What a noble field, kind reader, for thy experimental philosophy and speculations, for thy learning, for thy perseverance, for thy kind-heartedness, for everything that is great and good within thee!

The accidental traveller who has journeyed on from Stabroek to the rock Saba, and from thence to the banks of the Essequibo, in pursuit of other things, as he told thee at the beginning, with but an indifferent interpreter to talk to, no friend to converse with, and totally unfit for that which he wishes thee to do, can merely mark the outlines of the path he has trodden, or tell thee the sounds he has heard, or faintly describe what he has seen in the environs of his resting-places; but if this be enough to induce thee to undertake the journey, and give the world a description of it, he will be amply satisfied.

It will be two days and a half from the time of entering the path on the western bank of the Demerara till all be ready, and the canoe fairly afloat on the Essequibo. The new rigging it, and putting every little thing to rights and in its proper place, cannot well be done in less than a day.

After being night and day in the forest impervious to the sun and moon's rays, the sudden transition to light has a fine heart-cheering effect. Welcome as a lost friend, the solar beam makes the frame rejoice, and with it a thousand enlivening thoughts rush at once on the soul, and disperse, as a vapour, every sad and sorrowful idea, which the deep gloom had helped to collect there. In coming out of the woods, you see the western bank of the Essequibo before you, low and flat. Here the river is two-thirds as broad as the Demerara at Stabroek.

To the northward there is a hill higher than any in the Demerara; and in the south-south-west quarter a mountain. It is far away, and appears like a bluish cloud in the horizon. There is not the least opening on either side. Hills, valleys, and lowlands, are all linked together by a chain of forest. Ascend the highest mountain, climb the loftiest tree, as far as the eye can extend, whichever way it directs itself, all is luxuriant and unbroken forest.

In about nine or ten hours from this, you get to an Indian habitation of three huts, on the point of an island. It is said that a Dutch post once stood here; but there is not the smallest vestige of it remaining, and, except that the trees appear younger than those on the other islands, which shows that the place has been cleared some time or other, there is no mark left by which you can conjecture that ever this was a post.

The many islands which you meet with in the way, enliven and change the scene, by the avenues which they make, which look like the mouths of other rivers, and break that long-extended sameness which is seen in the Demerara.

Proceeding onwards, you get to the falls and rapids. In the rainy season they are very tedious to pass, and often stop your course. In the dry season, by stepping from rock to rock, the Indians soon manage to get a canoe over them. But when the river is swollen, as it was in May, 1812, it is then a difficult task, and often a dangerous one too. At that time many of the islands were overflowed, the rocks covered, and the lower branches of the trees in the water. Sometimes the Indians were obliged to take everything out of the canoe, cut a passage through the branches, which hung over into the river, and then drag up the canoe by main force.

At one place, the falls form an oblique line quite across the river, impassable to the ascending canoe, and you are forced to have it dragged four or five hundred yards by land.

It will take you five days, from the Indian habitation on the point of the island, to where these falls and rapids terminate.

There are no huts in the way. You must bring your own cassava-bread along with you, hunt in the forest for your meat, and make the night's shelter for yourself.

Here is a noble range of hills, all covered with the finest trees, rising majestically one above the other, on the western bank, and presenting as rich a scene as ever the eye would wish to look on. Nothing in vegetable nature can be conceived more charming, grand, and luxuriant.

How the heart rejoices in viewing this beautiful landscape! when the sky is serene, the air cool, and the sun just sunk behind the mountain's top.

The hayawa-tree perfumes the woods around; pairs of scarlet aras are continually crossing the river. The maam sends forth its plaintive note, the wren chants its evening song. The caprimulgus wheels in busy flight around the canoe, while "whip-poor-will" sits on the broken stump near the water's edge, complaining as the shades of night set in.

A little before you pass the last of these rapids two immense rocks appear, nearly on the summit of one of the many hills which form this far-extending range where it begins to fall off gradually to the south.

They look like two ancient stately towers of some Gothic potentate, rearing their heads above the surrounding trees. What with their situation and their shape together, they strike the beholder with an idea of antiquated grandeur which he will never forget. He may travel far and near and see nothing like them. On looking at them through a glass, the summit of the southern one appeared crowned with bushes. The one to the north was quite bare. The Indians have it from their ancestors that they are the abode of an evil genius, and they pass in the river below with a reverential awe.

In about seven hours from these stupendous sons of the hill, you leave the Essequibo, and enter the river Apourapoura, which falls into it from the south. The Apourapoura is nearly one-third the size of the Demerara at Stabroek. For two days you see nothing but level ground, richly clothed in timber. You leave the Siparouni to the right hand, and on the third day come to a little hill. The Indians have cleared about an acre of ground on it, and erected a temporary shed. If it be not intended for provision-ground alone, perhaps the next white man who travels through these remote wilds will find an Indian settlement here.

Two days after leaving this, you get to a rising ground on the western bank, where stands a single hut; and about half a mile in the forest there are a few more; some of them square, and some round with spiral roofs.

Here the fish called pacou is very plentiful: it is perhaps the fattest and most delicious fish in Guiana. It does not take the hook, but the Indians decoy it to the surface of the water by means of the seeds of the crabwood-tree, and then shoot it with an arrow.

You are now within the borders of Macoushia, inhabited by a different tribe of people, called Macoushi Indians; uncommonly dexterous in the use of the blowpipe, and famous for their skill in preparing the deadly vegetable poison commonly called wourali.

It is from this country that those beautiful paroquets named kessi-kessi are procured. Here the crystal mountains are found; and here the three different species of the ara are seen in great abundance. Here, too, grows the tree from which the gum-elastic is got; it is large, and as tall as any in the forest. The wood has much the appearance of sycamore. The gum is contained in the bark: when that is cut through it oozes out very freely: it is quite white, and looks as rich as cream: it hardens almost immediately as it issues from the tree; so that it is very easy to collect a ball, by forming the juice into a globular shape as fast as it comes out; it becomes nearly black by being exposed to the air, and is real Indian rubber without undergoing any other process.

The elegant crested bird called cock of the rock, admirably described by Buffon, is a native of the woody mountains of Macoushia. In the daytime he retires amongst the darkest rocks, and only comes out to feed a little before sunrise, and at sunset; he is of a gloomy disposition, and, like the houtou, never associates with the other birds of the forest.

The Indians in the just-mentioned settlement seemed to depend more on the wourali-poison for killing their game than upon anything else. They had only one gun, and it appeared rusty and neglected; but their poisoned weapons were in fine order. Their blowpipes hung from the roof of the hut, carefully suspended by a silk-grass cord; and on taking a nearer view of them, no dust seemed to have collected there, nor had the spider spun the smallest web on them; which showed that they were in constant use. The quivers were close by them, with the jaw-bone of the fish pirai tied by a string to their brim, and a small wicker-basket of wild cotton, which hung down to the centre; they were nearly full of poisoned arrows. It was with difficulty these Indians could be persuaded to part with any of the wourali-poison, though a good price was offered for it; they gave me to understand that it was powder and shot to them, and very difficult to be procured.

On the second day after leaving the settlement, in passing along, the Indians show you a place where once a white man lived. His retiring so far from those of his own colour and acquaintance seemed to carry something extraordinary along with it, and raised a desire to know what could have induced him to do so. It seems he had been unsuccessful, and that his creditors had treated him with as little mercy as the strong generally show to the weak. Seeing his endeavours daily frustrated, and his best intentions of no avail, and fearing that when they had taken all he had they would probably take his liberty too, he thought the world would not be hard-hearted enough to condemn him for retiring from the evils which pressed so heavily on him, and which he had done all that an honest man could do to ward off. He left his creditors to talk of him as they thought fit, and bidding adieu for ever to the place in which he had once seen better times, he penetrated thus far into those remote and gloomy wilds, and ended his days here.

According to the new map of South America, Lake Parima, or the White Sea, ought to be within three or four days' walk from this place. On asking the Indians whether there was such a place or not, and describing that the water was fresh and good to drink, an old Indian, who appeared to be about sixty, said that there was such a place, and that he had been there. This information would have been satisfactory in some degree, had not the Indians carried the point a little too far. It is very large, said another Indian, and ships come to it. Now these unfortunate ships were the very things which were not wanted; had he kept them out, it might have done, but his introducing them was sadly against the lake. Thus you must either suppose that the old savage and his companion had a confused idea of the thing, and that probably the Lake Parima they talked of was the Amazons, not far from the city of Para, or that it was their intention to deceive you. You ought to be cautious in giving credit to their stories, otherwise you will be apt to be led astray.

Many a ridiculous thing concerning the interior of Guiana has been propagated and received as true, merely because six or seven Indians, questioned separately, have agreed in their narrative.

Ask those who live high up in the Demerara, and they will, every one of them, tell you that there is a nation of Indians with long tails; that they are very malicious, cruel and ill-natured; and that the Portuguese have been obliged to stop them off in a certain river, to prevent their depredations. They have also dreadful stories concerning a horrible beast, called the watermamma, which, when it happens to take a spite against a canoe, rises out of the river, and in the most unrelenting manner possible carries both canoe and Indians down to the bottom with it, and there destroys them. Ludicrous extravagances; pleasing to those fond of the marvellous, and excellent matter for a distempered brain.

The misinformed and timid court of policy in Demerara was made the dupe of a savage, who came down the Essequibo, and gave himself out as king of a mighty tribe. This naked wild man of the woods seemed to hold the said court in tolerable contempt, and demanded immense supplies, all which he got; and moreover, some time after, an invitation to come down the ensuing year for more, which he took care not to forget.

This noisy chieftain boasted so much of his dynasty and domain, that the Government was induced to send up an expedition into his territories to see if he had spoken the truth, and nothing but the truth. It appeared, however, that his palace was nothing but a hut, the monarch a needy savage, the heir-apparent nothing to inherit but his father's club and bow and arrows, and his officers of state wild and uncultivated as the forests through which they strayed.

There was nothing in the hut of this savage, saving the presents he had received from Government, but what was barely sufficient to support existence; nothing that indicated a power to collect a hostile force; nothing that showed the least progress towards civilisation. All was rude and barbarous in the extreme, expressive of the utmost poverty and a scanty population.

You may travel six or seven days without seeing a hut, and when you reach a settlement it seldom contains more than ten.

The further you advance into the interior the more you are convinced that it is thinly inhabited.

The day after passing the place where the white man lived you see a creek on the left hand, and shortly after the path to the open country. Here you drag the canoe up into the forest, and leave it there. Your baggage must now be carried by the Indians. The creek you passed in the river intersects the path to the next settlement: a large mora has fallen across it, and makes an excellent bridge. After walking an hour and a half you come to the edge of the forest, and a savanna unfolds itself to the view.

The finest park that England boasts falls far short of this delightful scene. There are about two thousand acres of grass, with here and there a clump of trees, and a few bushes and single trees scattered up and down by the hand of Nature. The ground is neither hilly nor level, but diversified with moderate rises and falls, so gently running into one another that the eye cannot distinguish where they begin, nor where they end, while the distant black rocks have the appearance of a herd at rest. Nearly in the middle there is an eminence, which falls off gradually on every side; and on this the Indians have erected their huts.

To the northward of them the forest forms a circle, as though it had been done by art; to the eastward it hangs in festoons; and to the south and west it rushes in abruptly, disclosing a new scene behind it at every step as you advance along.

This beautiful park of nature is quite surrounded by lofty hills, all arrayed in superbest garb of trees; some in the form of pyramids, others like sugar-loaves towering one above the other; some rounded off, and others as though they had lost their apex. Here two hills rise up in spiral summits, and the wooded line of communication betwixt them sinks so gradually that it forms a crescent; and there the ridges of others resemble the waves of an agitated sea. Beyond these appear others, and others past them; and others still farther on, till they can scarcely be distinguished from the clouds.

There are no sand-flies, nor bête-rouge, nor mosquitos in this pretty spot. The fire-flies during the night vie in numbers and brightness with the stars in the firmament above: the air is pure, and the north-east breeze blows a refreshing gale throughout the day. Here the white-crested maroudi, which is never found in the Demerara, is pretty plentiful; and here grows the tree which produces the moran, sometimes called balsam capivi.

Your route lies south from this place; and at the extremity of the savanna you enter the forest, and journey along a winding path at the foot of a hill. There is no habitation within this day's walk. The traveller, as usual, must sleep in the forest. The path is not so good the following day. The hills over which it lies are rocky, steep, and rugged, and the spaces betwixt them swampy, and mostly knee-deep in water. After eight hours' walk you find two or three Indian huts, surrounded by the forest; and in little more than half an hour from these you come to ten or twelve others, where you pass the night. They are prettily situated at the entrance into a savanna. The eastern and western hills are still covered with wood; but on looking to the south-west quarter you perceive it begins to die away. In those forests you may find plenty of the trees which yield the sweet-smelling resin called acaiari, and which, when pounded and burnt on charcoal, gives a delightful fragrance.

From hence you proceed, in a south-west direction, through a long swampy savanna. Some of the hills which border on it have nothing but a thin coarse grass and huge stones on them; others, quite wooded; others with their summits crowned, and their base quite bare; and others, again, with their summits bare, and their base in thickest wood.

Half of this day's march is in water, nearly up to the knees. There are four creeks to pass; one of them has a fallen tree across it. You must make your own bridge across the other three. Probably, were the truth known, these apparently four creeks are only the meanders of one.

The jabiru, the largest bird in Guiana, feeds in the marshy savanna through which you have just passed. He is wary and shy, and will not allow you to get within gun-shot of him.

You sleep this night in the forest, and reach an Indian settlement about three o'clock the next evening, after walking one-third of the way through wet and miry ground.

But, bad as the walking is through it, it is easier than where you cross over the bare hills, where you have to tread on sharp stones, most of them lying edgewise.

The ground gone over these two last days seems condemned to perpetual solitude and silence. There was not one four-footed animal to be seen, nor even the marks of one. It would have been as silent as midnight, and all as still and unmoved as a monument had not the jabiru in the marsh, and a few vultures soaring over the mountain's top, shown that it was not quite deserted by animated nature. There were no insects, except one kind of fly about one-fourth the size of the common house-fly. It bit cruelly, and was much more tormenting than the mosquito on the sea-coast.

This seems to be the native country of the arrowroot. Wherever you passed through a patch of wood in a low situation, there you found it growing luxuriantly.

The Indian place you are now at is not the proper place to have come to in order to reach the Portuguese frontiers. You have advanced too much to the westward. But there was no alternative. The ground twixt you and another small settlement (which was the right place to have gone to) was overflowed; and thus, instead of proceeding southward, you were obliged to wind along the foot of the western hills, quite out of your way.

But the grand landscape this place affords makes you ample amends for the time you have spent in reaching it. It would require great descriptive powers to give a proper idea of the situation these people have chosen for their dwelling.

The hill they are on is steep and high, and full of immense rocks. The huts are not all in one place, but dispersed wherever they have found a place level enough for a lodgment. Before you ascend the hill you see at intervals an acre or two of wood, then an open space, with a few huts on it, then wood again, and then an open space, and so on, till the intervening of the western hills, higher and steeper still, and crowded with trees of the loveliest shades, closes the enchanting scene.

At the base of this hill stretches an immense plain, which appears to the eye, on this elevated spot, as level as a bowling-green. The mountains on the other side are piled one upon the other in romantic forms, and gradually retire, till they are indiscernible from the clouds in which they are involved. To the south-south-west this far-extending plain is lost in the horizon. The trees on it, which look like islands on the ocean, add greatly to the beauty of the landscape; while the rivulet's course is marked out by the æta-trees which follow its meanders.

Not being able to pursue the direct course from hence to the next Indian habitation on account of the floods of water which fall at this time of the year, you take a circuit westerly along the mountain's foot.

At last a large and deep creek stops your progress: it is wide and rapid, and its banks very steep. There is neither curial nor canoe, nor purple-heart tree in the neighbourhood to make a wood-skin to carry you over, so that you are obliged to swim across; and by the time you have formed a kind of raft, composed of boughs of trees and coarse grass, to ferry over your luggage, the day will be too far spent to think of proceeding. You must be very cautious before you venture to swim across this creek, for the alligators are numerous, and near twenty feet long. On the present occasion the Indians took uncommon precautions lest they should be devoured by this cruel and voracious reptile. They cut long sticks, and examined closely the side of the creek for half a mile above and below the place where it was to be crossed; and as soon as the boldest had swum over, he did the same on the other side, and then all followed.

After passing the night on the opposite bank, which is well wooded, it is a brisk walk of nine hours before you reach four Indian huts, on a rising ground a few hundred paces from a little brook, whose banks are covered over with coucourito and æta trees.

This is the place you ought to have come to two days ago had the water permitted you. In crossing the plain at the most advantageous place you are above ankle-deep in water for three hours; the remainder of the way is dry, the ground gently rising. As the lower parts of this spacious plain put on somewhat the appearance of a lake during the periodical rains, it is not improbable but that this is the place which hath given rise to the supposed existence of the famed Lake Parima, or El Dorado; but this is mere conjecture.

A few deer are feeding on the coarse rough grass of this far-extending plain; they keep at a distance from you, and are continually on the look-out.

The spur-winged plover, and a species of the curlew, black, with a white bar across the wings, nearly as large again as the scarlet curlew on the sea-coast, frequently rise before you. Here, too, the Moscovy duck is numerous; and large flocks of two other kinds wheel round you as you pass on, but keep out of gun-shot. The milk-white egrets, and jabirus, are distinguished at a great distance; and in the æta and coucourito trees you may observe flocks of scarlet and blue aras feeding on the seeds.

It is to these trees that the largest sort of toucan resorts. He is remarkable by a large black spot on the point of his fine yellow bill. He is very scarce in Demerara, and never seen except near the sea-coast.

The ants' nests have a singular appearance on this plain. They are in vast abundance on those parts of it free from water, and are formed of an exceeding hard yellow clay. They rise eight or ten feet from the ground in a spiral form, impenetrable to the rain, and strong enough to defy the severest tornado.

The wourali-poison, procured in these last-mentioned huts, seemed very good, and proved afterwards to be very strong.

There are now no more Indian settlements betwixt you and the Portuguese frontiers. If you wish to visit their fort, it would be advisable to send an Indian with a letter from hence, and wait his return. On the present occasion a very fortunate circumstance occurred. The Portuguese commander had sent some Indians and soldiers to build a canoe, not far from this settlement; they had just finished it, and those who did not stay with it had stopped here on their return.

The soldier who commanded the rest, said he durst not, upon any account, convoy a stranger to the fort; but he added, as there were two canoes, one of them might be despatched with a letter, and then we could proceed slowly on in the other.

About three hours from this settlement there is a river called Pirarara; and here the soldiers had left their canoes while they were making the new one. From the Pirarara you get into the river Maou, and then into the Tacatou; and just where the Tacatou falls into the Rio Branco, there stands the Portuguese frontier fort, called Fort St. Joachim. From the time of embarking in the river Pirarara, it takes you four days before you reach this fort.

There was nothing very remarkable in passing down these rivers. It is an open country, producing a coarse grass, and interspersed with clumps of trees. The banks have some wood on them, but it appears stunted and crooked, like that on the bleak hills in England.

The tapir frequently plunged into the river; he was by no means shy, and it was easy to get a shot at him on land. The kessi-kessi paroquets were in great abundance; and the fine scarlet ara innumerable in the coucourito trees at a distance from the river's bank. In the Tacatou was seen the troupiale. It was charming to hear the sweet and plaintive notes of this pretty songster of the wilds. The Portuguese call it the nightingale of Guiana.

Towards the close of the fourth evening, the canoe, which had been sent on with a letter, met us with the commander's answer. During its absence, the nights had been cold and stormy, the rain had fallen in torrents, the days cloudy, and there was no sun to dry the wet hammocks. Exposed thus, day and night, to the chilling blast and pelting shower, strength of constitution at last failed, and a severe fever came on. The commander's answer was very polite. He remarked, he regretted much to say that he had received orders to allow no stranger to enter the frontier, and this being the case, he hoped I would not consider him as uncivil. "However," continued he, "I have ordered the soldier to land you at a certain distance from the fort, where we can consult together."

We had now arrived at the place, and the canoe which brought the letter returned to the fort, to tell the commander I had fallen sick.

The sun had not risen above an hour the morning after when the Portuguese officer came to the spot where we had landed the preceding evening. He was tall and spare, and appeared to be from fifty to fifty-five years old; and, though thirty years of service under an equatorial sun had burnt and shrivelled up his face, still there was something in it so inexpressibly affable and kind, that it set you immediately at your ease. He came close up to the hammock, and taking hold of my wrist to feel the pulse, "I am sorry, sir," said he, "to see that the fever has taken such hold of you. You shall go directly with me," continued he, "to the fort; and though we have no doctor there, I trust," added he, "we shall soon bring you about again. The orders I have received forbidding the admission of strangers were never intended to be put in force against a sick English gentleman."

As the canoe was proceeding slowly down the river towards the fort, the commander asked, with much more interest than a question in ordinary conversation is asked, where was I on the night of the first of May? On telling him that I was at an Indian settlement a little below the great fall in the Demerara, and that a strange and sudden noise had alarmed all the Indians, he said the same astonishing noise had roused every man in Fort St. Joachim, and that they remained under arms till morning. He observed that he had been quite at a loss to form any idea what could have caused the noise; but now learning that the same noise had been heard at the same time far away from the Rio Branco, it struck him there must have been an earthquake somewhere or other.

Good nourishment and rest, and the unwearied attention and kindness of the Portuguese commander, stopped the progress of the fever, and enabled me to walk about in six days.

Fort St. Joachim was built about five-and-forty years ago, under the apprehension, it is said, that the Spaniards were coming from the Rio Negro to settle there. It has been much neglected; the floods of water have carried away the gate, and destroyed the wall on each side of it; but the present commander is putting it into thorough repair. When finished, it will mount six nine- and six twelve-pounders.

In a straight line with the fort, and within a few yards of the river, stand the commander's house, the barracks, the chapel, the father confessor's house, and two others, all at little intervals from each other; and these are the only buildings at Fort St. Joachim. The neighbouring extensive plains afford good pasturage for a fine breed of cattle, and the Portuguese make enough of butter and cheese for their own consumption.

On asking the old officer if there were such a place as Lake Parima, or El Dorado, he replied, he looked upon it as imaginary altogether. "I have been above forty years," added he, "in Portuguese Guiana, but have never yet met with anybody who has seen the lake."

So much for Lake Parima, or El Dorado, or the White Sea. Its existence at best seems doubtful; some affirm that there is such a place, and others deny it.

> "Grammatici certant, et adhuc sub judice lis est."

Having now reached the Portuguese inland frontier, and collected a sufficient quantity of the wourali-poison, nothing remains but to give a brief account of its composition, its effects, its uses, and its supposed antidotes.

It has been already remarked that in the extensive wilds of Demerara and Essequibo, far away from any European settlement, there is a tribe of Indians who are known by the name of Macoushi.

Though the wourali-poison is used by all the South American savages betwixt the Amazons and the Oroonoque, still this tribe makes it stronger than any of the rest. The Indians in the vicinity of the Rio Negro are aware of this, and come to the Macoushi country to purchase it.

Much has been said concerning this fatal and extraordinary poison. Some have affirmed that its effects are almost instantaneous, provided the minutest particle of it mixes with the blood; and others again have maintained that it is not strong enough to kill an animal of the size and strength of a man. The first have erred by lending a too willing ear to the marvellous, and believing assertions without sufficient proof. The following short story points out the necessity of a cautious examination:—

One day, on asking an Indian if he thought the poison would kill a man, he replied that they always go to battle with it; that he was standing by when an Indian was shot with a poisoned arrow, and that he expired almost immediately. Not wishing to dispute this apparently satisfactory information, the subject was dropped. However, about an hour after, having purposely asked him in what part of the body the said Indian was wounded, he answered without hesitation that the arrow entered betwixt his shoulders, and passed quite through his heart. Was it the weapon, or the strength of the poison, that brought on immediate dissolution in this case? Of course the weapon.

The second have been misled by disappointment, caused by neglect in keeping the poisoned arrows, or by not knowing how to use them, or by trying inferior poison. If the arrows are not kept dry, the poison loses its strength; and in wet or damp weather it turns mouldy, and becomes quite soft. In shooting an arrow in this state, upon examining the place where it has entered, it will be observed that, though the arrow has penetrated deep into the flesh, still by far the greatest part of the poison has shrunk back, and thus, instead of entering with the arrow, it has remained collected at the mouth of the wound. In this case the arrow might as well not have been poisoned. Probably, it was to this that a gentleman, some time ago, owed his disappointment, when he tried the poison on a horse in the town of Stabroek, the capital of Demerara; the horse never betrayed the least symptom of being affected by it.

Wishful to obtain the best information concerning this poison, and as repeated inquiries, in lieu of dissipating the surrounding shade, did but tend more and more to darken the little light that existed, I determined to penetrate into the country where the poisonous ingredients grow, where this pernicious composition is prepared, and where it is constantly used. Success attended the adventure; and the information acquired made amends for one hundred and twenty days passed in the solitudes of Guiana, and afforded a balm to the wounds and bruises which every traveller must expect to receive who wanders through a thorny and obstructed path.

Thou must not, courteous reader, expect a dissertation on the manner in which the wourali-poison operates on the system; a treatise has been already written on the subject; and, after all, there is probably still reason to doubt. It is supposed to affect the nervous system, and thus destroy the vital functions; it is also said to be perfectly harmless, provided it does not touch the blood. However, this is certain, when a sufficient quantity of it enters the blood, death is the inevitable consequence; but there is no alteration in the colour of the blood, and both the blood and flesh may be eaten with safety.

All that thou wilt find here is a concise, unadorned account of the wourali-poison. It may be of service to thee some time or other, shouldst thou ever travel through the wilds where it is used. Neither attribute to cruelty, nor to a want of feeling for the sufferings of the inferior animals, the ensuing experiments. The larger animals were destroyed in order to have proof positive of the strength of a poison which hath hitherto been doubted: and the smaller ones were killed with the hope of substantiating that which has commonly been supposed to be an antidote.

It makes a pitying heart ache to see a poor creature in distress and pain; and too often has the compassionate traveller occasion to heave a sigh as he journeys on. However, here, though the kind-hearted will be sorry to read of an unoffending animal doomed to death in order to satisfy a doubt, still it will be a relief to know that the victim was not tortured. The wourali-poison destroys life's action so gently, that the victim appears to be in no pain whatever; and probably, were the truth known, it feels none, saving the momentary smart at the time the arrow enters.

A day or two before the Macoushi Indian prepares his poison, he goes into the forest in quest of the ingredients. A vine grows in these wilds, which is called wourali. It is from this that the poison takes its name, and it is the principal ingredient. When he has procured enough of this, he digs up a root of a very bitter taste, ties them together, and then looks about for two kinds of bulbous plants, which contain a green and glutinous juice. He fills a little quake, which he carries on his back, with the stalks of these; and, lastly, ranges up and down till he finds two species of ants. One of them is very large and black, and so venomous that its sting produces a fever; it is most commonly to be met with on the ground. The other is a little red ant, which stings like a nettle, and generally has its nest under the leaf of a shrub. After obtaining these, he has no more need to range the forest.

A quantity of the strongest Indian pepper is used; but this he has already planted round his hut. The pounded fangs of the labarri snake, and those of the counacouchi, are likewise added. These he commonly has in store; for when he kills a snake, he generally extracts the fangs, and keeps them by him.

Having thus found the necessary ingredients, he scrapes the wourali vine and bitter root into thin shavings, and puts them into a kind of colander made of leaves; this he holds over an earthen pot, and pours water on the shavings: the liquor which comes through has the appearance of coffee. When a sufficient quantity has been procured, the shavings are thrown aside. He then bruises the bulbous stalks and squeezes a proportionate quantity of their juice through his hands into the pot. Lastly, the snakes' fangs, ants, and pepper are bruised, and thrown into it. It is then placed on a slow fire, and as it boils more of the juice of the wourali is added, according as it may be found necessary, and the scum is taken off with a leaf: it remains on the fire till reduced to a thick syrup of a deep brown colour. As soon as it has arrived at this state, a few arrows are poisoned with it to try its strength. If it answers the expectations, it is poured out into a calabash, or little pot of Indian manufacture, which is carefully covered with a couple of leaves, and over them a piece of deer's skin, tied round with a cord. They keep it in the most dry part of the hut; and from time to time suspend it over the fire, to counteract the effects of dampness.

The act of preparing this poison is not considered as a common one: the savage may shape his bow, fasten the barb on the point of his arrow, and make his other implements of destruction, either lying in his hammock, or in the midst of his family; but if he has to prepare the wourali-poison, many precautions are supposed to be necessary.

The women and young girls are not allowed to be present, lest the yabahou, or evil spirit, should do them harm. The shed under which it has been boiled has been pronounced polluted and abandoned ever after. He who makes the poison must eat nothing that morning, and must continue fasting as long as the operation lasts. The pot in which it is boiled must be a new one, and must never have held anything before, otherwise the poison would be deficient in strength: add to this that the operator must take particular care not to expose himself to the vapour which arises from it while on the fire.

Though this and other precautions are taken, such as frequently washing the face and hands, still the Indians think that it affects the health; and the operator either is, or, what is more probable, supposes himself to be, sick for some days after.

Thus it appears that the making the wourali-poison is considered as a gloomy and mysterious operation; and it would seem that they imagine it affects others as well as him who boils it; for an Indian agreed one evening to make some for me, but the next morning he declined having anything to do with it, alleging that his wife was with child!

Here it might be asked, are all the ingredients just mentioned necessary, in order to produce the wourali-poison? Though our opinions and conjectures may militate against the absolute necessity of some of them, still it would be hardly fair to pronounce them added by the hand of superstition, till proof positive can he obtained.

We might argue on the subject, and, by bringing forward instances of Indian superstition, draw our conclusion by inference, and still remain in doubt on this head. You know superstition to be the offspring of ignorance, and of course that it takes up its abode amongst the rudest tribes of uncivilised man. It even too often resides with man in his more enlightened state.

The Augustan age furnishes numerous examples. A bone snatched from the jaws of a fasting bitch, and a feather from the wing of a night-owl,— "ossa ab ore rapta jejunæ canis, plumamque nocturnæ strigis,"—were necessary for Canidia's incantations. And in aftertimes, Parson Evans, the Welshman, was treated most ungenteelly by an enraged spirit, solely because he had forgotten a fumigation in his witch-work. If, then, enlightened man lets his better sense give way, and believes, or allows himself to be persuaded, that certain substances and actions, in reality of no avail, possess a virtue which renders them useful in producing the wished-for effect; may not the wild, untaught, unenlightened savage of Guiana, add an ingredient which, on account of the harm it does him, he fancies may be useful in the perfection of his poison, though, in fact, it be of no use at all? If a bone snatched from the jaws of a fasting bitch be thought necessary in incantation; or if witchcraft have recourse to the raiment of the owl, because it resorts to the tombs and mausoleums of the dead, and wails and hovers about at the time that the rest of animated nature sleeps; certainly the savage may imagine that the ants, whose sting causes a fever, and the teeth of the labarri and counacouchi snakes, which convey death in a very short space of time, are essentially necessary in the composition of his poison; and being once impressed with this idea, he will add them every time he makes the poison, and transmit the absolute use of them to his posterity. The question to be answered seems not to be if it is natural for the Indians to mix these ingredients, but if they are essential to make the poison.

So much for the preparing of this vegetable essence—terrible importer of death into whatever animal it enters. Let us now see how it is used; let us examine the weapons which bear it to its destination, and take a view of the poor victim, from the time he receives his wound till death comes to his relief.

When a native of Macoushia goes in quest of feathered game or other birds, he seldom carries his bow and arrows. It is the blow-pipe he then uses. This extraordinary tube of death is, perhaps, one of the greatest natural curiosities of Guiana. It is not found in the country of the Macoushi. Those Indians tell you that it grows to the south-west of them, in the wilds which extend betwixt them and the Rio Negro. The reed must grow to an amazing length, as the part the Indians use is from ten to eleven feet long, and no tapering can be perceived in it, one end being as thick as the other. It is of a bright yellow colour, perfectly smooth both inside and out. It grows hollow; nor is there the least appearance of a knot or joint throughout the whole extent. The natives call it ourah. This, of itself, is too slender to answer the end of a blow-pipe; but there is a species of palma, larger and stronger, and common in Guiana, and this the Indians make use of as a case, in which they put the ourah. It is brown, susceptible of a fine polish, and appears as if it had joints five or six inches from each other. It is called samourah, and the pulp inside is easily extracted, by steeping it for a few days in water.

Thus the ourah and samourah, one within the other, form the blow-pipe of Guiana. The end which is applied to the mouth is tied round with a small silk-grass cord, to prevent its splitting; and the other end, which is apt to strike against the ground, is secured by the seed of the acuero fruit, cut horizontally through the middle, with a hole made in the end, through which is put the extremity of the blow-pipe. It is fastened on with string on the outside, and the inside is filled up with wild bees'-wax.

The arrow is from nine to ten inches long. It is made out of the leaf of a species of palm-tree, called coucourito, hard and brittle, and pointed as sharp as a needle. About an inch of the pointed end is poisoned. The other end is burnt to make it still harder, and wild cotton is put round it for about an inch and a half. It requires considerable practice to put on this cotton well. It must just be large enough to fit the hollow of the tube, and taper off to nothing downwards. They tie it on with a thread of the silk-grass, to prevent its slipping off the arrow.

The Indians have shown ingenuity in making a quiver to hold the arrows. It will contain from five to six hundred. It is generally from twelve to fourteen inches long, and in shape resembles a dice-box used at backgammon. The inside is prettily done in basket-work, with wood not unlike bamboo, and the outside has a coat of wax. The cover is all of one piece, formed out of the skin of the tapir. Round the centre there is fastened a loop, large enough to admit the arm and shoulder, from which it hangs when used. To the rim is tied a little bunch of silk-grass, and half of the jawbone of the fish called pirai, with which the Indian scrapes the point of his arrow.

Before he puts the arrows into the quiver, he links them together by two strings of cotton, one string at each end, and then folds them round a stick, which is nearly the length of the quiver. The end of the stick, which is uppermost, is guarded by two little pieces of wood crosswise, with a hoop round their extremities, which appears something like a wheel; and this saves the hand from being wounded when the quiver is reversed in order to let the bunch of arrows drop out.

There is also attached to the quiver a little kind of basket, to hold the wild cotton, which is put on the blunt end of the arrow. With a quiver of poisoned arrows slung over his shoulder, and with his blowpipe in his hand, in the same position as a soldier carries his musket, see the Macoushi Indian advancing towards the forest in quest of powises, maroudis, waracabas, and other feathered game.

These generally sit high up in the tall and tufted trees, but still are not out of the Indian's reach; for his blowpipe, at its greatest elevation, will send an arrow three hundred feet. Silent as midnight he steals under them, and so cautiously does he tread the ground that the fallen leaves rustle not beneath his feet. His ears are open to the least sound, while his eye, keen as that of the lynx, is employed in finding out the game in the thickest shade. Often he imitates their cry, and decoys them from tree to tree, till they are within range of his tube. Then taking a poisoned arrow from his quiver, he puts it in the blowpipe, and collects his breath for the fatal puff.

About two feet from the end through which he blows there are fastened two teeth of the acouri, and these serve him for a sight. Silent and swift the arrow flies, and seldom fails to pierce the object at which it is sent. Sometimes the wounded bird remains in the same tree where it was shot, and in three minutes falls down at the Indian's feet. Should he take wing, his flight is of short duration, and the Indian, following the direction he has gone, is sure to find him dead.

It is natural to imagine that, when a slight wound only is inflicted, the game will make its escape. Far otherwise; the wourali-poison almost instantaneously mixes with blood or water, so that if you wet your finger, and dash it along the poisoned arrow in the quickest manner possible, you are sure to carry off some of the poison. Though three minutes generally elapse before the convulsions come on in the wounded bird, still a stupor evidently takes place sooner, and this stupor manifests itself by an apparent unwillingness in the bird to move. This was very visible in a dying fowl.

Having procured a healthy full-grown one, a short piece of a poisoned blowpipe arrow was broken off and run up into its thigh, as near as possible betwixt the skin and the flesh, in order that it might not be incommoded by the wound. For the first minute it walked about, but walked very slowly, and did not appear the least agitated. During the second minute it stood still, and began to peck the ground; and ere half another had elapsed, it frequently opened and shut its mouth. The tail had now dropped, and the wings almost touched the ground. By the termination of the third minute, it had sat down, scarce able to support its head, which nodded, and then recovered itself, and then nodded again, lower and lower every time, like that of a weary traveller slumbering in an erect position; the eyes alternately open and shut. The fourth minute brought on convulsions, and life and the fifth terminated together.

The flesh of the game is not in the least injured by the poison, nor does it appear to corrupt sooner than that killed by the gun or knife. The body of this fowl was kept for sixteen hours, in a climate damp and rainy, and within seven degrees of the equator; at the end of which time it had contracted no bad smell whatever, and there were no symptoms of putrefaction, saving that, just round the wound, the flesh appeared somewhat discoloured.

The Indian, on his return home, carefully suspends his blowpipe from the top of his spiral roof; seldom placing it in an oblique position, lest it should receive a cast.

Here let the blowpipe remain suspended, while you take a view of the arms which are made to slay the larger beasts of the forest.

When the Indian intends to chase the peccari, or surprise the deer, or rouse the tapir from his marshy retreat, he carries his bow and arrows, which are very different from the weapons already described.

The bow is generally from six to seven feet long, and strung with a cord, spun out of the silk-grass. The forests of Guiana furnish many species of hard wood, tough and elastic, out of which beautiful and excellent bows are formed.

The arrows are from four to five feet in length, made of a yellow reed without a knot or joint. It is found in great plenty up and down throughout Guiana. A piece of hard wood, about nine inches long, is inserted into the end of the reed, and fastened with cotton well waxed. A square hole, an inch deep, is then made in the end of this piece of hard wood, done tight round with cotton to keep it from splitting. Into this square hole is fitted a spike of coucourite wood, poisoned, and which may be kept there, or taken out at pleasure. A joint of bamboo, about as thick as your finger, is fitted on over the poisoned spike, to prevent accidents and defend it from the rain, and is taken off when the arrow is about to be used. Lastly, two feathers are fastened on the other end of the reed to steady it in its flight.

Besides his bow and arrows, the Indian carries a little box made of bamboo, which holds a dozen or fifteen poisoned spikes six inches long. They are poisoned in the following manner: a small piece of wood is dipped in the poison, and with this they give the spike a first coat. It is then exposed to the sun or fire. After it is dry, it receives another coat, and is then dried again; after this a third coat, and sometimes a fourth.

They take great care to put the poison on thicker at the middle than at the sides, by which means the spike retains the shape of a two-edged sword. It is rather a tedious operation to make one of these arrows complete; and as the Indian is not famed for industry, except when pressed by hunger, he has hit upon a plan of preserving his arrows which deserves notice.

About a quarter of an inch above the part where the coucourite spike is fixed into the square hole, he cuts it half through; and thus, when it has entered the animal, the weight of the arrow causes it to break off there, by which means the arrow falls to the ground uninjured; so that, should this be the only arrow he happens to have with him, and should another shot immediately occur, he has only to take another poisoned spike out of his little bamboo box, fit it on his arrow, and send it to its destination.

Thus armed with deadly poison, and hungry as the hyæna, he ranges through the forest in quest of the wild beast's track. No hound can act a surer part. Without clothes to fetter him, or shoes to bind his feet, he observes the footsteps of the game, where an European eye could not discern the smallest vestige. He pursues it through all its turns and windings with astonishing perseverance, and success generally crowns his efforts. The animal, after receiving the poisoned arrow, seldom retreats two hundred paces before it drops.

In passing overland from the Essequibo to the Demerara we fell in with a herd of wild hogs. Though encumbered with baggage, and fatigued with a hard day's walk, an Indian got his bow ready; and let fly a poisoned arrow at one of them. It entered the cheek-bone and broke off. The wild hog was found quite dead about one hundred and seventy paces from the place where he had been shot. He afforded us an excellent and wholesome supper.

Thus the savage of Guiana, independent of the common weapons of destruction, has it in his power to prepare a poison, by which he can generally ensure to himself a supply of animal food; and the food so destroyed imbibes no deleterious qualities. Nature has been bountiful to him. She has not only ordered poisonous herbs and roots to grow in the unbounded forests through which he strays, but has also furnished an excellent reed for his arrows, and another, still more singular, for his blowpipe; and planted trees of an amazing hard, tough, and elastic texture, out of which he forms his bows. And in order that nothing might be wanting, she has superadded a tree which yields him a fine wax, and disseminated up and down, a plant not unlike that of the pineapple, which affords him capital bowstrings.

Having now followed the Indian in the chase, and described the poison, let us take a nearer view of its action, and observe a large animal expiring under the weight of its baneful virulence.

Many have doubted the strength of the wourali poison. Should they ever by chance read what follows, probably their doubts on that score will be settled for ever.

In the former experiment on the hog some faint resistance on the part of nature was observed, as if existence struggled for superiority; but in the following instance of the sloth life sank in death without the least apparent contention, without a cry, without a struggle, and without a groan. This was an ai, or three-toed sloth. It was in the possession of a gentleman who was collecting curiosities. He wished to have it killed, in order to preserve the skin, and the wourali-poison was resorted to as the easiest death.

Of all animals, not even the toad and tortoise excepted, this poor ill-formed creature is the most tenacious of life. It exists long after it has received wounds which would have destroyed any other animal; and it may be said, on seeing a mortally wounded sloth, that life disputes with death every inch of flesh in its body.

The ai was wounded in the leg, and put down on the floor, about two feet from the table; it contrived to reach the leg of the table, and fastened himself on it, as if wishful to ascend. But this was its last advancing step: life was ebbing fast, though imperceptibly; nor could this singular production of nature, which has been formed of a texture to resist death in a thousand shapes, make any stand against the wourali-poison.

First, one fore-leg let go its hold, and dropped down motionless by its side; the other gradually did the same. The fore-legs having now lost their strength, the sloth slowly doubled its body, and placed its head betwixt its hind-legs, which still adhered to the table; but when the poison had affected these also it sank to the ground, but sank so gently, that you could not distinguish the movement from an ordinary motion; and had you been ignorant that it was wounded with a poisoned arrow, you would never have suspected that it was dying. Its mouth was shut, nor had any froth or saliva collected there.

There was no *subsultus tendinum*, or any visible alteration in its breathing. During the tenth minute from the time it was wounded it stirred, and that was all; and the minute after life's last spark went out. From the time the poison began to operate, you would have conjectured that sleep was overpowering it, and you would have exclaimed, "Pressitque jacentem, dulcis et alta quies, placidæque simillima morti."

There are now two positive proofs of the effect of this fatal poison, viz., the death of the hog, and that of the sloth. But still these animals were nothing remarkable for size; and the strength of the poison in large animals might yet be doubted, were it not for what follows.

A large well-fed ox, from nine hundred to a thousand pounds' weight, was tied to a stake by a rope sufficiently long to allow him to move to and fro. Having no large coucourito spikes at hand, it was judged necessary, on account of his superior size, to put three wild-hog arrows into him; one was sent into each thigh just above the hock, in order to avoid wounding a vital part, and the third was shot transversely into the extremity of the nostril.

The poison seemed to take effect in four minutes. Conscious as though he would fall, the ox set himself firmly on his legs, and remained quite still in the same place till about the fourteenth minute, when he smelled the ground, and appeared as if inclined to walk. He advanced a pace or two, staggered, and fell, and remained extended on his side with his head on the ground. His eye, a few minutes ago so bright and lively, now became fixed and dim, and though you put your hand close to it, as if to give him a blow there, he never closed his eyelid.

His legs were convulsed, and his head from time to time started involuntarily; but he never showed the least desire to raise it from the ground; he breathed hard, and emitted foam from his mouth. The startings, or *subsultus tendinum*, now became gradually weaker and weaker; his hinder parts were fixed in death; and in a minute or two more his head and fore-legs ceased to stir.

Nothing now remained to show that life was still within him, except that his heart faintly beat and fluttered at intervals. In five-and-twenty minutes from the time of his being wounded he was quite dead. His flesh was very sweet and savoury at dinner.

On taking a retrospective view of the two different kinds of poisoned arrows, and the animals destroyed by them, it would appear that the quantity of poison must be proportioned to the animal, and thus those probably labour under an error who imagine that the smallest particle of it introduced into the blood has almost instantaneous effects.

Make an estimate of the difference in size betwixt the fowl and the ox, and then weigh a sufficient quantity of poison for a blowpipe arrow with which the fowl was killed, and weigh also enough poison for three wild-hog arrows which destroyed the ox, and it will appear that the fowl received much more poison in proportion than the ox. Hence the cause why the fowl died in five minutes and the ox in five-and-twenty.

Indeed, were it the case that the smallest particle of it introduced into the blood has almost instantaneous effects, the Indian would not find it necessary to make the large arrow; that of the blowpipe is much easier made and requires less poison.

And now for the antidotes, or rather the supposed antidotes. The Indians tell you, that if the wounded animal be held for a considerable time up to the mouth in water, the poison will not prove fatal; also that the juice of the sugar-cane poured down the throat will counteract the effects of it. These antidotes were fairly tried upon full-grown healthy fowls, but they all died, as though no steps had been taken to preserve their lives. Rum was recommended and given to another, but with as little success.

It is supposed by some that wind introduced into the lungs by means of a small pair of bellows would revive the poisoned patient, provided the operation be continued for a sufficient length of time. It may be so; but this is a difficult and a tedious mode of cure, and he who is wounded in the forest far away from his friends, or in the hut of the savages, stands but a poor chance of being saved by it.

Had the Indians a sure antidote, it is likely they would carry it about with them, or resort to it immediately after being wounded, if at hand; and their confidence in its efficacy would greatly diminish the horror they betray when you point a poisoned arrow at them.

One day, while we were eating a red monkey, erroneously called the baboon of Demerara, an Arowack Indian told an affecting story of what happened to a comrade of his. He was present at his death. As it did not interest this Indian in any point to tell a falsehood it is very probable that his account was a true one. If so, it appears that there is no certain antidote, or at least an antidote that could be resorted to in a case of urgent need; for the Indian gave up all thoughts of life as soon as he was wounded.

The Arowack Indian said it was but four years ago that he and his companion were ranging in the forest in quest of game. His companion took a poisoned arrow, and sent it at a red monkey in a tree above him. It was nearly a perpendicular shot. The arrow missed the monkey, and in the descent struck him in the arm a little above the elbow. He was convinced it was all over with him. "I shall never," said he to his companion in a faltering voice, and looking at his bow as he said it, "I shall never," said he, "bend this bow again." And having said that he took off his little bamboo poison-box, which hung across his shoulder, and putting it, together with his bow and arrows, on the ground, he laid himself down close by them, bid his companion farewell, and never spoke more.

He who is unfortunate enough to be wounded by a poisoned arrow from Macoushia had better not depend upon the common antidotes for a cure. Many who have been in Guiana will recommend immediate immersion in water, or to take the juice of the sugar-cane, or to fill the mouth full of salt; and, they recommend these antidotes because they have got them from the Indians. But were you to ask them if they ever saw these antidotes used with success, it is ten to one their answer would be in the negative.

Wherefore let him reject these antidotes as unprofitable, and of no avail. He has got an active and a deadly foe within him, which, like Shakespeare's fell Sergeant Death, is strict in his arrest, and will allow him but little time—very, very little time. In a few minutes he will be numbered with the dead. Life ought, if possible, to be preserved, be the expense ever so great. Should the part affected admit of it, let a ligature be tied tight round the wound, and have immediate recourse to the knife:—

> "Continuo, culpam ferro compesce priusquam,
> Dira per infaustum serpant contagia corpus."

And now, kind reader, it is time to bid thee farewell. The two ends proposed have been obtained. The Portuguese inland frontier fort has been reached, and the Macoushi wourali-poison acquired. The account of this excursion through the interior of Guiana has been submitted to thy perusal, in order to induce thy abler genius to undertake a more extensive one. If any difficulties have arisen, or fevers come on, they have been caused by the periodical rains, which fall in torrents as the sun approaches the tropic of Cancer. In dry weather there would be no difficulties or sickness.

Amongst the many satisfactory conclusions which thou wouldst be able to draw during the journey, there is one which, perhaps, would please thee not a little, and that is with regard to dogs. Many a time, no doubt, thou hast heard it hotly disputed that dogs existed in Guiana previously to the arrival of the Spaniards in those parts. Whatever the Spaniards introduced, and which bore no resemblance to anything the Indians had been accustomed to see, retains its Spanish name to this day.

Thus the Warrow, the Arowack, the Acoway, the Macoushi, and Carib tribes, call a hat, "sombrero;" shirt, or any kind of cloth, "camisa;" a shoe, "zapato;" a letter, "carta;" a fowl, "gallina;" gunpowder, "colvora" (Spanish, "polvora"); ammunition, "bala;" a cow, "vaca;" and a dog, "perro."

This argues strongly against the existence of dogs in Guiana before it was discovered by the Spaniards, and probably may be of use to thee in thy next canine dispute.

In a political point of view this country presents a large field for speculation. A few years ago there was but little inducement for any Englishman to explore the interior of these rich and fine colonies, as the British Government did not consider them worth holding at the peace of Amiens. Since that period their mother-country has been blotted out from the list of nations, and America has unfolded a new sheet of politics. On one side the crown of Braganza, attacked by an ambitious chieftain, has fled from the palace of its ancestors, and now seems fixed on the banks of the Janeiro. Cayenne has yielded to its arms. La Plata has raised the standard of independence, and thinks itself sufficiently strong to obtain a government of its own. On the other side, the Caraccas are in open revolt, and should Santa Fé join them in good earnest they may form a powerful association.

Thus, on each side of the *ci-devant* Dutch Guiana, most unexpected and astonishing changes have taken place. Will they raise or lower it in the scale of estimation at the Court of St. James's? Will they be of benefit to these grand and extensive colonies?—colonies enjoying perpetual summer—colonies of the richest soil—colonies containing within themselves everything necessary for their support—colonies, in fine, so varied in their quality and situation, as to be capable of bringing to perfection every tropical production; and only want the support of government, and an enlightened governor, to render them as fine as the finest portions of the equatorial regions. Kind reader, fare thee well.

**LETTER TO THE PORTUGUESE COMMANDER.**

MUY SEÑOR,

Como no tengo el honor, de ser conocido de VM. lo pienso mejor, y mas decoroso, quedarme aqui, hastaque huviere recibido su respuesta. Haviendo caminado hasta la chozo, adonde estoi, no quisiere volverme, antes de haver visto la fortaleza de los Portugueses; y pido licencia de VM. para que me adelante. Honradissimos son mis motivos, ni tengo proyecto ninguno, o de comercio, o de la soldadesca, no siendo yo, o comerciante, o oficial. Hidalgo catolico soy, de hacienda in Ynglatierra, y muchos años de mi vida he pasado en caminar. Ultimamente, de Demeraria vengo, la qual dexé el 5 dia de Abril, para ver este hermoso pais, y coger unas curiosidades, especialmente, el veneno, que se llama wourali. Las mas recentes noticias que tenian en Demeraria, antes de mi salida, eran medias tristes, medias alegres. Tristes digo, viendo que Valencia ha caido en poder del enemigo comun, y el General Blake, y sus valientes tropas, quedan prisioneros de guerra. Alegres, al contrario, porque Milord Wellington se ha apoderado de Ciudad Rodrigo. A pesar de la caida de Valencia, parece claro al mundo, que las cosas del enemigo estan andando de pejor a pejor cada dia. Nosotros debemos dar gracias al Altissimo, por haver sido servido dexarnos castigar ultimamente a los robadores de sus santas Yglesias. Se vera VM. que yo no escribo Portugues ni aun lo hablo, pero, haviendo aprendido el Castellano, no nos faltará medio de communicar y tener conversacion. Ruego se escuse esta carta escrita sin tinta, porque un Indio dexo caer mi tintero y quebrose. Dios le dé a VM. muchos años de salud. Entretanto, tengo el honor de ser

Su mas obedeciente servidor,
CARLOS WATERTON.

## REMARKS.

"Incertus, quo fata ferant, ubi sistera detur."

Kind and gentle reader, if the journey in quest of the wourali-poison has engaged thy attention, probably thou mayst recollect that the traveller took leave of thee at Fort St. Joachim, on the Rio Branco. Shouldst thou wish to know what befell him afterwards, excuse the following uninteresting narrative.

Having had a return of fever, and aware that the farther he advanced into these wild and lonely regions the less would be the chance of regaining his health, he gave up all idea of proceeding onwards, and went slowly back towards the Demerara nearly by the same route he had come.

On descending the falls in the Essequibo, which form an oblique line quite across the river, it was resolved to push through them, the downward stream being in the canoe's favour. At a little distance from the place a large tree had fallen into the river, and in the meantime the canoe was lashed to one of its branches.

The roaring of the water was dreadful; it foamed and dashed over the rocks with a tremendous spray, like breakers on a lee-shore, threatening destruction to whatever approached it. You would have thought, by the confusion it caused in the river, and the whirlpools it made, that Scylla and Charybdis, and their whole progeny, had left the Mediterranean, and come and settled here. The channel was barely twelve feet wide, and the torrent in rushing down formed transverse furrows, which showed how near the rocks were to the surface.

Nothing could surpass the skill of the Indian who steered the canoe. He looked steadfastly at it, then at the rocks, then cast an eye on the channel, and then looked at the canoe again. It was in vain to speak. The sound was lost in the roar of waters; but his eye showed that he had already passed it in imagination. He held up his paddle in a position, as much as to say, that he would keep exactly amid channel; and then made a sign to cut the bush rope that held the canoe to the fallen tree. The canoe drove down the torrent with inconceivable rapidity. It did not touch the rocks once all the way. The Indian proved to a nicety "medio tutissimus ibis."

Shortly after this it rained almost day and night, the lightning flashing incessantly, and the roar of thunder awful beyond expression.

The fever returned, and pressed so heavy on him, that to all appearance his last day's march was over. However, it abated; his spirits rallied, and he marched again; and after delays and inconveniences he reached the house of his worthy friend Mr. Edmonstone, in Mibiri Creek, which falls into the

Demerara. No words of his can do justice to the hospitality of that gentleman, whose repeated encounters with the hostile negroes in the forest have been publicly rewarded, and will be remembered in the colony for years to come.

Here he learned that an eruption had taken place in St. Vincent's; and thus the noise heard in the night of the first of May, which had caused such terror amongst the Indians, and made the garrison at Fort St. Joachim remain under arms the rest of the night, is accounted for.

After experiencing every kindness and attention from Mr. Edmonstone, he sailed for Granada, and from thence to St. Thomas's, a few days before poor Captain Peake lost his life on his own quarter-deck, bravely fighting for his country on the coast of Guiana.

At St. Thomas's they show you a tower, a little distance from the town, which, they say, formerly belonged to a buccaneer chieftain. Probably the fury of besiegers has reduced it to its present dismantled state. What still remains of it bears testimony of its former strength, and may brave the attack of time for centuries. You cannot view its ruins without calling to mind the exploits of those fierce and hardy hunters, long the terror of the western world. While you admire their undaunted courage, you lament that it was often stained with cruelty; while you extol their scrupulous justice to each other, you will find a want of it towards the rest of mankind. Often possessed of enormous wealth, often in extreme poverty, often triumphant on the ocean, and often forced to fly to the forests, their life was an ever-changing scene of advance and retreat, of glory and disorder, of luxury and famine. Spain treated them as outlaws and pirates, while other European Powers publicly disowned them. They, on the other hand, maintained that injustice on the part of Spain first forced them to take up arms in self-defence; and that, whilst they kept inviolable the laws which they had framed for their own common benefit and protection, they had a right to consider as foes those who treated them as outlaws. Under this impression they drew the sword, and rushed on as though in lawful war, and divided the spoils of victory in the scale of justice.

After leaving St. Thomas's a severe tertian ague every now and then kept putting the traveller in mind that his shattered frame, "starting and shivering in the inconstant blast, meagre and pale—the ghost of what it was"—wanted repairs. Three years elapsed after arriving in England before the ague took its final leave of him.

During that time several experiments were made with the wourali-poison. In London an ass was inoculated with it, and died in twelve minutes. The poison was inserted into the leg of another, round which a bandage had been previously tied a little above the place where the wourali was introduced. He walked about as usual, and ate his food as though all were right. After an hour had elapsed the bandage was untied, and ten minutes after death overtook him.

A she-ass received the wourali-poison in the shoulder, and died apparently in ten minutes. An incision was then made in its windpipe, and through it the lungs were regularly inflated for two hours with a pair of bellows. Suspended animation returned. The ass held up her head, and looked around; but the inflating being discontinued, she sank once more in apparent death. The artificial breathing was immediately recommenced, and continued without intermission for two hours; this saved the ass from final dissolution. She rose up, and walked about; she seemed neither in agitation nor in pain. The wound, through which the poison entered, was healed without difficulty. Her constitution, however, was so severely affected that it was long a doubt if ever she would be well again. She looked lean and sickly for above a year, but began to mend the spring after, and by Midsummer became fat and frisky.

The kind-hearted reader will rejoice on learning that Earl Percy, pitying her misfortunes, sent her down from London to Walton Hall, near Wakefield. There she goes by the name of Wouralia. Wouralia shall be sheltered from the wintry storm; and when summer comes she shall feed in the finest pasture. No burden shall be placed upon her, and she shall end her days in peace.

For three revolving autumns the ague-beaten wanderer never saw, without a sigh, the swallow bend her flight towards warmer regions. He wished to go too, but could not; for sickness had enfeebled him, and prudence pointed out the folly of roving again too soon across the northern tropic. To be sure, the Continent was now open, and change of air might prove beneficial; but there was nothing very tempting in a trip across the Channel, and as for a tour through England—England has long ceased to be the land for adventures. Indeed, when good King Arthur reappears to claim his crown he will find things strangely altered here; and may we not look for his coming? for there is written upon his gravestone:—

>"Hic jacet Arturus, Rex quondam Rexque futurus,"
>"Here Arthur lies, who formerly
>Was king—and king again to be."

Don Quixote was always of opinion that this famous king did not die, but that he was changed into a raven by enchantment, and that the English are momentarily expecting his return. Be this as it may, it is certain that when he reigned here all was harmony and joy. The browsing herds passed from vale to vale, the swains sang from the bluebell-teeming groves, and nymphs, with eglantine and roses in their neatly-braided hair, went hand in hand to the flowery mead to weave garlands for their lambkins. If by chance some rude uncivil fellow dared to molest them, or attempted to throw thorns in their path, there was sure to be a knight-errant not far off ready to rush forward in their defence. But alas! in these degenerate days it is not so. Should a harmless cottage-maid wander out of the highway to pluck a primrose or two in the neighbouring field the haughty owner sternly bids her retire; and if a pitying swain hasten to escort her back, he is perhaps seized by the gaunt house-dog ere he reach her.

Æneas's route on the other side of Styx could not have been much worse than this, though by his account, when he got back to earth, it appears that he had fallen in with "Bellua Lernæ, horrendum stridens, flammisque, armata Chimæra."

Moreover, he had a sibyl to guide his steps; and as such a conductress nowadays could not be got for love nor money, it was judged most prudent to refrain from sauntering through this land of freedom, and wait with patience the return of health. At last this long-looked-for, ever-welcome stranger came.

# SECOND JOURNEY.

In the year 1816, two days before the vernal equinox, I sailed from Liverpool for Pernambuco, in the southern hemisphere, on the coast of Brazil. There is little at this time of the year in the European part of the Atlantic to engage the attention of the naturalist. As you go down the Channel you see a few divers and gannets. The middle-sized gulls, with a black spot at the end of the wings, attend you a little way into the Bay of Biscay. When it blows a hard gale of wind the stormy petrel makes its appearance. While the sea runs mountains high, and every wave threatens destruction to the labouring vessel, this little harbinger of storms is seen enjoying itself, on rapid pinion, up and down the roaring billows. When the storm is over it appears no more. It is known to every English sailor by the name of Mother Carey's chicken. It must have been hatched in Æolus's cave, amongst a clutch of squalls and tempests; for whenever they get out upon the ocean it always contrives to be of the party.

Though the calms and storms and adverse winds in these latitudes are vexatious, still, when you reach the trade winds, you are amply repaid for all disappointments and inconveniences. The trade winds prevail about thirty degrees on each side of the equator. This part of the ocean may be called the Elysian Fields of Neptune's empire; and the torrid zone, notwithstanding Ovid's remark, "non est habitabilis æstu," is rendered healthy and pleasant by these gently-blowing breezes. The ship glides smoothly on, and you soon find yourself within the northern tropic. When you are on it, Cancer is just over your head, and betwixt him and Capricorn is the high road of the Zodiac, forty-seven degrees wide, famous for Phaeton's misadventure. His father begged and entreated him not to take it into his head to drive parallel to the five zones, but to mind and keep on the turnpike which runs obliquely across the equator. "There you will distinctly see," said he, "the ruts of my chariot wheels, 'manifesta rotæ vestigia cernes.' But," added he, "even suppose you keep on it, and avoid the byroads, nevertheless, my dear boy, believe me, you will be most sadly put to your shifts; 'ardua prima via est,' the first part of the road is confoundedly steep! 'ultima via prona est,' and after that it is all down hill. Moreover, 'per insidias iter est, formasque ferarum,' the road is full of nooses and bull-dogs, 'Hæmoniosque arcus,' and spring guns, 'sævaque circuitu, curvantem brachia longo, Scorpio,' and steel traps of uncommon size and shape." These were nothing in the eyes of Phaeton; go he would, so off he set, full speed, four-in-hand. He had a tough drive of it; and after doing a prodigious deal of mischief, very luckily for the world, he got thrown out of the box, and tumbled into the river Po.

Some of our modern bloods have been shallow enough to try to ape this poor empty-headed coachman, on a little scale, making London their Zodiac. Well for them if tradesmen's bills, and other trivial perplexities, have not caused them to be thrown into the King's Bench.

The productions of the torrid zone are uncommonly grand. Its plains, its swamps, its savannas, and forests abound with the largest serpents and wild beasts; and its trees are the habitation of the most beautiful of the feathered race. While the traveller in the Old World is astonished at the elephant, the tiger, the lion, and the rhinoceros, he who wanders through the torrid regions of the New is lost in admiration at the cotingas, the toucans, the humming-birds, and aras.

The ocean, likewise, swarms with curiosities. Probably the flying-fish may be considered as one of the most singular. This little scaled inhabitant of water and air seems to have been more favoured than the rest of its finny brethren. It can rise out of the waves, and on wing visit the domain of the birds.

After flying two or three hundred yards, the intense heat of the sun has dried its pellucid wings, and it is obliged to wet them in order to continue its flight. It just drops into the ocean for a moment, and then rises again and flies on; and then descends to remoisten them, and then up again into the air; thus passing its life, sometimes wet, sometimes dry, sometimes in sunshine, and sometimes in the pale moon's nightly beam, as pleasure dictates, or as need requires. The additional assistance of wings is not thrown away upon it. It has full occupation both for fins and wings, as its life is in perpetual danger.

The bonito and albicore chase it day and night; but the dolphin is its worst and swiftest foe. If it escape into the air, the dolphin pushes on with proportional velocity beneath, and is ready to snap it up the moment it descends to wet its wings.

You will often see above one hundred of these little marine aërial fugitives on the wing at once. They appear to use every exertion to prolong their flight, but vain are all their efforts; for when the last drop of water on their wings is dried up, their flight is at an end, and they must drop into the ocean. Some are instantly devoured by their merciless pursuer, part escape by swimming, and others get out again as quick as possible, and trust once more to their wings.

It often happens that this unfortunate little creature, after alternate dips and flights, finding all its exertions of no avail, at last drops on board the vessel, verifying the old remark—

"Incidit in Scyllam, cupiens vitare Charybdim."

There, stunned by the fall, it beats the deck with its tail, and dies. When eating it, you would take it for a fresh herring. The largest measure from fourteen to fifteen inches in length. The dolphin, after pursuing it to the ship, sometimes forfeits his own life.

In days of yore, the musician used to play in softest, sweetest strain, and then take an airing amongst the dolphins; "inter delphinas Arion." But nowadays, our tars have quite capsized the custom; and instead of riding ashore on the dolphin, they invited the dolphin aboard. While he is darting and playing around the vessel, a sailor goes out to the spritsailyard-arm, and with a long staff, leaded at one end, and armed at the other with five barbed spikes, he heaves it at him. If successful in his aim, there is a fresh mess for all hands. The dying dolphin affords a superb and brilliant sight:

"Mille trahit moriens, adverso sole colores."

All the colours of the rainbow pass and repass in rapid succession over his body, till the dark hand of death closes the scene.

From the Cape de Verd Islands to the coast of Brazil you see several different kinds of gulls, which probably are bred in the island of St. Paul. Sometimes the large bird called the frigate pelican soars majestically over the vessel, and the tropic-bird comes near enough to let you have a fair view of the long feathers in his tail. On the line when it is calm sharks of a tremendous size make their appearance. They are descried from the ship by means of the dorsal fin, which is above the water.

On entering the Bay of Pernambuco, the frigate pelican is seen watching the shoals of fish from a prodigious height. It seldom descends without a successful attack on its numerous prey below.

As you approach the shore the view is charming. The hills are clothed with wood, gradually rising towards the interior, none of them of any considerable height. A singular reef of rocks runs parallel to the coast, and forms the harbour of Pernambuco. The vessels are moored betwixt it and the town, safe from every storm. You enter the harbour through a very narrow passage, close by a fort built on the reef. The hill of Olinda, studded with houses and convents, is on your right hand, and an island, thickly planted with cocoa-nut trees, adds considerably to the scene on your left. There are two strong forts on the isthmus, betwixt Olinda and Pernambuco, and a pillar midway to aid the pilot.

Pernambuco probably contains upwards of fifty thousand souls. It stands on a flat, and is divided into three parts—a peninsula, an island, and the continent. Though within a few degrees of the line, its climate is remarkably salubrious, and rendered almost temperate by the refreshing sea-breeze. Had art and judgment contributed their portion to its natural advantages, Pernambuco at this day would have been a stately ornament to the coast of Brazil. On viewing it, it will strike you that every one has built his house entirely for himself, and deprived public convenience of the little claim she had a right to put in. You would wish that this city, so famous for its harbour, so happy in its climate, and so well situated for commerce, could have risen under the flag of Dido, in lieu of that of Braganza.

As you walk down the streets, the appearance of the homes is not much in their favour. Some of them are very high, and some very low; some newly whitewashed, and others stained and mouldy and neglected, as though they had no owner.

The balconies, too, are of a dark and gloomy appearance. They are not, in general, open, as in most tropical cities, but grated like a farmer's dairy window, though somewhat closer.

There is a lamentable want of cleanliness in the streets. The impurities from the houses, and the accumulation of litter from the beasts of burden, are unpleasant sights to the passing stranger. He laments the want of a police as he goes along; and when the wind begins to blow, his nose and eyes are too often exposed to a cloud of very unsavoury dust.

When you view the port of Pernambuco, full of ships of all nations; when you know that the richest commodities of Europe, Africa, and Asia are brought to it; when you see immense quantities of cotton, dye-wood, and the choicest fruits pouring into the town, you are apt to wonder at the little attention these people pay to the common comforts which one always expects to find in a large and opulent city. However, if the inhabitants are satisfied, there is nothing more to be said. Should they ever be convinced that inconveniences exist, and that nuisances are too frequent, the remedy is in their own hands. At present, certainly, they seem perfectly regardless of them; and the Captain-General of Pernambuco walks through the streets with as apparent content and composure as an English statesman would proceed down Charing Cross. Custom reconciles everything. In a week or two the stranger himself begins to feel less the things which annoyed him so much upon his first arrival, and after a few months' residence he thinks no more about them, while he is partaking of the hospitality and enjoying the elegance and splendour within doors in this great city.

Close by the riverside stands what is called the Palace of the Captain-General of Pernambuco. Its form and appearance altogether strike the traveller that it was never intended for the use it is at present put to.

Reader, throw a veil over thy recollection for a little while, and forget the cruel, unjust, and unmerited censures thou hast heard against an unoffending order. This palace was once the Jesuits' college, and originally built by those charitable fathers. Ask the aged and respectable inhabitants of Pernambuco, and they will tell thee that the destruction of the Society of Jesus was a terrible disaster to the public, and its consequences severely felt to the present day.

When Pombal took the reins of power into his own hands, virtue and learning beamed bright within the college walls. Public catechism to the children, and religious instruction to all, flowed daily from the mouths of its venerable priests.

They were loved, revered, and respected throughout the whole town. The illuminating philosophers of the day had sworn to exterminate Christian knowledge, and the college of Pernambuco was doomed to founder in the general storm. To the long-lasting sorrow and disgrace of Portugal, the philosophers blinded her king and flattered her Prime Minister. Pombal was exactly the tool these sappers of every public and private virtue wanted. He had the naked sword of power in his own hand, and his heart was as hard as flint. Ho struck a mortal blow, and the Society of Jesus throughout the Portuguese dominions, was no more.

One morning all the fathers of the college in Pernambuco, some of them very old and feeble, were suddenly ordered into the refectory. They had notice beforehand of the fatal storm, in pity from the governor, but not one of them abandoned his charge. They had done their duty, and had nothing to fear. They bowed with resignation to the will of Heaven. As soon as they had all reached the refectory, they were there locked up, and never more did they see their rooms, their friends, their scholars, or acquaintance. In the dead of the following night, a strong guard of soldiers literally drove them through the streets to the water's edge. They were then conveyed in boats aboard a ship, and steered for Bahia. Those who survived the barbarous treatment they experienced from Pombal's creatures were at last ordered to Lisbon. The college of Pernambuco was plundered, and some time after an elephant was kept there.

Thus the arbitrary hand of power, in one night, smote and swept away the sciences; to which succeeded the low vulgar buffoonery of a showman. Virgil and Cicero made way for a wild beast from Angola! and now a guard is on duty at the very gate where, in times long past, the poor were daily fed!!!

Trust not, kind reader, to the envious remarks which their enemies have scattered far and near; believe not the stories of those who have had a hand in the sad tragedy. Go to Brazil, and see with thine own eyes the effect of Pombal's short-sighted policy. There vice reigns triumphant, and learning is at its lowest ebb. Neither is this to be wondered at. Destroy the compass, and will the vessel find her far-distant port? Will the flock keep together, and escape the wolves, after the shepherds are all slain? The Brazilians were told that public education would go on just as usual. They might have asked Government, who so able to instruct our youth as those whose knowledge is proverbial? who so fit as those who enjoy our entire confidence? who so worthy, as those whose lives are irreproachable?

They soon found that those who succeeded the fathers of the Society of Jesus had neither their manner nor their abilities. They had not made the instruction of youth their particular study. Moreover, they entered on the field after a defeat, where the officers had all been slain; where the plan of the campaign was lost; where all was in sorrow and dismay. No exertions of theirs could rally the dispersed, or skill prevent the fatal consequences. At the present day the seminary of Olinda, in comparison with the former Jesuits' college, is only as the waning moon's beam to the sun's meridian splendour.

When you visit the places where those learned fathers once flourished, and see with your own eyes the evils their dissolution has caused; when you hear the inhabitants telling you how good, how clever, how charitable they were—what will you think of our poet laureate for calling them, in his "History of Brazil," "Missioners, whose zeal the most fanatical was directed by the coolest policy"?

Was it *fanatical* to renounce the honours and comforts of this transitory life, in order to gain eternal glory in the next, by denying themselves, and taking up the cross? Was it *fanatical* to preach salvation to innumerable wild hordes of Americans, to clothe the naked, to encourage the repenting sinner, to aid the dying Christian? The fathers of the Society of Jesus did all this. And for this their zeal is pronounced to be the most fanatical, directed by the coolest policy. It will puzzle many a clear brain to comprehend how it is possible, in the nature of things, that *zeal* the most *fanatical* should be directed by the *coolest policy*. Ah, Mr. Laureate, Mr. Laureate, that "quidlibet audendi" of yours may now and then gild the poet, at the same time that it makes the historian cut a sorry figure!

Could Father Nobrega rise from the tomb, he would thus address you:— "Ungrateful Englishman, you have drawn a great part of your information from the writings of the Society of Jesus, and in return you attempt to stain its character by telling your countrymen that 'we taught the idolatry we believed!' In speaking of me, you say, it was my happy fortune to be stationed in a country where *none* but the good principles of my order were called into action. Ungenerous laureate, the narrow policy of the times has kept your countrymen in the dark with regard to the true character of the Society of Jesus; and you draw the bandage still tighter over their eyes by a malicious insinuation. I lived, and taught, and died in Brazil, where you state that *none* but the good principles of my order were called into action, and still, in most absolute contradiction to this, you remark we believed the *idolatry* we taught in Brazil. Thus we brought none but good principles into action, and still taught idolatry!

"Again, you state there is no individual to whose talents Brazil is so greatly and permanently indebted as mine, and that I must be regarded as the founder of that system so successfully pursued by the Jesuits in Paraguay; a system productive of as much good as is compatible with pious fraud. Thus you make me, at one and the same time, a teacher of none but good principles, and a teacher of idolatry, and a believer in idolatry, and still the founder of a system for which Brazil is greatly and permanently indebted to me, though, by-the-bye, the system was only productive of as much good as is compatible with pious fraud!

"What means all this? After reading such incomparable nonsense, should your countrymen wish to be properly informed concerning the Society of Jesus, there are in England documents enough to show that the system of the Jesuits was a system of Christian charity towards their fellow-creatures, administered in a manner which human prudence judged best calculated to ensure success; and that the idolatry which you uncharitably affirm they taught was really and truly the very same faith which the Catholic Church taught for centuries in England, which she still teaches to those who wish to hear her, and which she will continue to teach pure and unspotted, till time shall be no more."

The environs of Pernambuco are very pretty. You see country houses in all directions, and the appearance of here and there a sugar plantation enriches the scenery. Palm-trees, cocoa-nut trees, orange and lemon groves, and all the different fruits peculiar to Brazil, are here in the greatest abundance.

At Olinda there is a national botanical garden; it wants space, produce, and improvement. The forests which are several leagues off, abound with birds, beasts, insects, and serpents. Besides a brilliant plumage, many of the birds have a very fine song. The troupiale, noted for its rich colours, sings delightfully in the environs of Pernambuco. The red-headed finch, larger than the European sparrow, pours forth a sweet and varied strain, in company with two species of wrens, a little before daylight. There are also several species of the thrush, which have a song somewhat different from that of the European thrush; and two species of the linnet, whose strain is so soft and sweet that it dooms them to captivity in the houses. A bird, called here *sangre do buey* (blood of the ox), cannot fail to engage your attention: he is of the passerine tribe, and very common about the houses; the wings and tail are black, and every other part of the body a flaming red. In Guiana there is a species exactly the same as this in shape, note, and economy, but differing in colour, its whole body being like black velvet; on its breast a tinge of red appears through the black. Thus nature has ordered this little tangara to put on mourning to the north of the line, and wears scarlet to the south of it.

For three months in the year the environs of Pernambuco are animated beyond description. From November to March the weather is particularly fine; then it is that rich and poor, young and old, foreigners and natives, all issue from the city to enjoy the country, till Lent approaches, when back they hie them. Villages and hamlets, where nothing before but rags were seen, now shine in all the elegance of dress; every house, every room, every shed become eligible places for those whom nothing but extreme necessity could have forced to live there a few weeks ago: some join in the merry dance, others saunter up and down the orange groves; and towards evening the roads become a moving scene of silk and jewels. The gaming-tables have constant visitors; there, thousands are daily and nightly lost and won; parties even sit down to try their luck round the outside of the door as well as in the room:—

> "Vestibulum ante ipsum primisque in faucibus aulæ
> Luctus et ultrices, posuere sedilia curæ."

About six or seven miles from Pernambuco stands a pretty little village called Monteiro; the river runs close by it, and its rural beauties seem to surpass all others in the neighbourhood; there the Captain-General of Pernambuco resides during this time of merriment and joy.

The traveller who allots a portion of his time to peep at his fellow-creatures in their relaxations, and accustoms himself to read their several little histories in their looks and gestures as he goes musing on, may have full occupation for an hour or two every day at this season amid the variegated scenes around the pretty village of Monteiro. In the evening groups sitting at the door, he may sometimes see with a sigh how wealth and the prince's favour cause a booby to pass for a Solon, and be reverenced as such, while perhaps a poor neglected Camoëns stands silent at a distance, awed by the dazzling glare of wealth and power. Retired from the public road he may see poor Maria sitting under a palm-tree, with her elbow in her lap and her head leaning on one side within her hand, weeping over her forbidden banns. And as he moves on, "with wandering step and slow," he may hear a broken-hearted nymph ask her faithless swain,

> "How could you say my face was fair,
>   And yet that face forsake;
> How could you win my virgin heart,
>   Yet leave that heart to break?"

One afternoon, in an unfrequented part not far from Monteiro, these adventures were near being brought to a speedy and a final close: six or seven blackbirds, with a white spot betwixt the shoulders, were making a noise, and passing to and fro on the lower branches of a tree in an abandoned, weed-grown, orange orchard. In the long grass underneath the tree, apparently a pale green grasshopper was fluttering, as though it had got entangled in it. When you once fancy that the thing you are looking at is really what you take it for, the more you look at it the more you are convinced it is so. In the present case, this was a grasshopper beyond all doubt, and nothing more remained to be done but to wait in patience till it had settled, in order that you might run no risk of breaking its legs in attempting to lay hold of it while it was fluttering—it still kept fluttering; and having quietly approached it, intended to make sure of it—behold, the head of a large rattlesnake appeared in the grass close by: an instantaneous spring backwards prevented fatal consequences. What had been taken for a grasshopper was, in fact, the elevated rattle of the snake in the act of announcing that he was quite prepared, though unwilling, to make a sure and deadly spring. He shortly after passed slowly from under the orange-tree to the neighbouring wood on the side of a hill: as he moved over a place bare of grass and weeds, he appeared to be about eight feet long; it was he who had engaged the attention of the birds, and made them heedless of danger from another quarter: they flew away on his retiring; one alone left his little life in the air, destined to become a specimen, mute and motionless, for the inspection of the curious in a far distant clime.

It was now the rainy season, the birds were moulting; fifty-eight specimens of the handsomest of them in the neighbourhood of Pernambuco had been collected, and it was time to proceed elsewhere. The conveyance to the interior was by horses; and this mode, together with the heavy rains, would expose preserved specimens to almost certain damage. The journey to Maranham by land would take at least forty days. The route was not wild enough to engage the attention of an explorer, or civilised enough to afford common comforts to a traveller. By sea there were no opportunities, except slave ships. As the transporting poor negroes from port to port for sale pays well in Brazil, the ships' decks are crowded with them. This would not do.

Excuse here, benevolent reader, a small tribute of gratitude to an Irish family, whose urbanity and goodness have long gained it the esteem and respect of all ranks in Pernambuco. The kindness and attention I received from Dennis Kearney, Esq., and his amiable lady, will be remembered with gratitude to my dying day.

After wishing farewell to this hospitable family, I embarked on board a Portuguese brig, with poor accommodation, for Cayenne in Guiana. The most eligible bedroom was the top of a hen-coop on deck. Even here, an unsavoury little beast, called bug, was neither shy nor deficient in appetite.

The Portuguese seamen are famed for catching fish. One evening, under the line, four sharks made their appearance in the wake of the vessel. The sailors caught them all.

On the fourteenth day after leaving Pernambuco, the brig cast anchor off the island of Cayenne. The entrance is beautiful. To windward, not far off, there are two bold wooded islands, called the Father and Mother; and near them are others, their children, smaller, though as beautiful as their parents. Another is seen a long way to leeward of the family, and seems as if it had strayed from home, and cannot find its way back. The French call it "l'enfant perdu." As you pass the islands, the stately hills on the main, ornamented with ever-verdant foliage, show you that this is by far the sublimest scenery on the sea-coast, from the Amazons to the Oroonoque. On casting your eye towards Dutch Guiana, you will see that the mountains become unconnected, and few in number, and long before you reach Surinam the Atlantic wave washes a flat and muddy shore.

Considerably to windward of Cayenne, and about twelve leagues from land, stands a stately and towering rock, called the Constable. As nothing grows on it to tempt greedy and aspiring man to claim it as his own, the sea-fowl rest and raise their offspring there. The bird called the frigate is ever soaring round its rugged summit. Hither the phaeton bends his rapid flight, and flocks of rosy flamingos here defy the fowler's cunning. All along the coast, opposite the Constable, and indeed on every uncultivated part of it to windward and leeward, are seen innumerable quantities of snow-white egrets, scarlet curlews, spoonbills, and flamingos.

Cayenne is capable of being a noble and productive colony. At present it is thought to be the poorest on the coast of Guiana. Its estates are too much separated one from the other by immense tracts of forest; and the revolutionary war, like a cold eastern wind, has chilled their zeal and blasted their best expectations.

The clove-tree, the cinnamon, pepper, and nutmeg, and many other choice spices and fruits of the eastern and Asiatic regions, produce abundantly in Cayenne.

The town itself is prettily laid out, and was once well fortified. They tell you it might easily have been defended against the invading force of the two united nations; but Victor Hugues, its governor, ordered the tri-coloured flag to be struck; and ever since that day the standard of Braganza has waved on the ramparts of Cayenne.

He who has received humiliations from the hand of this haughty, iron-hearted governor, may see him now in Cayenne, stripped of all his revolutionary honours, broken down and ruined, and under arrest in his own house. He has four accomplished daughters, respected by the whole town. Towards the close of day, when the sun's rays are no longer oppressive, these much-pitied ladies are seen walking up and down the balcony with their aged parent, trying, by their kind and filial attention, to remove the settled gloom from his too guilty brow.

This was not the time for a traveller to enjoy Cayenne. The hospitality of the inhabitants was the same as ever, but they had lost their wonted gaiety in public, and the stranger might read in their countenances, as the recollection of recent humiliations and misfortunes every now and then kept breaking in upon them, that they were still in sorrow for their fallen country: the victorious hostile cannon of Waterloo still sounded in their ears; their Emperor was a prisoner amongst the hideous rocks of St. Helena; and many a Frenchman who had fought and bled for France was now amongst them begging for a little support to prolong a life which would be forfeited on the parent soil. To add another handful to the cypress and wormwood already scattered amongst these polite colonists, they had just received orders from the court of Janeiro to put on deep mourning for six months, and half-mourning for as many more, on account of the death of the Queen of Portugal.

After a day's journey in the interior is the celebrated national plantation. This spot was judiciously chosen, for it is out of the reach of enemies' cruisers. It is called La Gabrielle. No plantation in the western world can vie with La Gabrielle. Its spices are of the choicest kind; its soil particularly favourable to them; its arrangements beautiful; and its directeur, Monsieur Martin, a botanist of first-rate abilities. This indefatigable naturalist ranged through the East, under a royal commission, in quest of botanical knowledge; and during his stay in the western regions has sent over to Europe from twenty to twenty-five thousand specimens in botany and zoology. La Gabrielle is on a far-extending range of woody hills. Figure to yourself a hill in the shape of a bowl reversed, with the buildings on the top of it, and you will have an idea of the appearance of La Gabrielle. You approach the house through a noble avenue, five hundred toises long, of the choicest tropical fruit-trees, planted with the greatest care and judgment; and should you chance to stray through it after sunset, when the clove-trees are in blossom, you would fancy yourself in the Idalian groves, or near the banks of the Nile, where they were burning the finest incense as the Queen of Egypt passed.

On La Gabrielle there are twenty-two thousand clove-trees in full bearing. They are planted thirty feet asunder. Their lower branches touch the ground. In general the trees are topped at five-and-twenty feet high; though you will see some here towering up above sixty. The black pepper, the cinnamon, and nutmeg are also in great abundance here, and very productive.

While the stranger views the spicy groves of La Gabrielle, and tastes the most delicious fruits which have originally been imported hither from all parts of the tropical world, he will thank the Government which has supported, and admire the talents of the gentleman who has raised to its present grandeur, this noble collection of useful fruits. There is a large nursery attached to La Gabrielle, where plants of all the different species are raised and distributed gratis to those colonists who wish to cultivate them.

Not far from the banks of the river Oyapoc, to windward of Cayenne, is a mountain which contains an immense cavern. Here the cock-of-the-rock is plentiful. He is about the size of a fantail-pigeon, his colour a bright orange, and his wings and tail appear as though fringed; his head is ornamented with a superb double-feathery crest, edged with purple. He passes the day amid gloomy damps and silence, and only issues out for food a short time at sunrise and sunset. He is of the gallinaceous tribe. The South-American Spaniards call him "gallo del Rio Negro" (cock of the Black River), and suppose that he is only to be met with in the vicinity of that far-inland stream; but he is common in the interior of Demerara, amongst the huge rocks in the forests of Macoushia; and he has been shot south of the line, in the captainship of Para.

The bird called by Buffon *grand gobemouche* has never been found in Demerara, although very common in Cayenne. He is not quite so large as the jackdaw, and is entirely black, except a large spot under the throat, which is a glossy purple.

You may easily sail from Cayenne to the river Surinam in two days. Its capital, Paramaribo, is handsome, rich, and populous: hitherto it has been considered by far the finest town in Guiana; but probably the time is not far off when the capital of Demerara may claim the prize of superiority.

You may enter a creek above Paramaribo, and travel through the interior of Surinam, till you come to the Nacari, which is close to the large river Coryntin. When you have passed this river, there is a good public road to New Amsterdam, the capital of Berbice.

On viewing New Amsterdam, it will immediately strike you that something or other has intervened to prevent its arriving at that state of wealth and consequence for which its original plan shows it was once intended. What has caused this stop in its progress to the rank of a fine and populous city remains for those to find out who are interested in it; certain it is that New Amsterdam has been languid for some years, and now the tide of commerce seems ebbing fast from the shores of Berbice.

Gay and blooming is the sister colony of Demerara. Perhaps, kind reader, thou hast not forgot that it was from Stabroek, the capital of Demerara, that the adventurer set out, some years ago, to reach the Portuguese frontier fort, and collected the wourali-poison. It was not intended, when this second sally was planned in England, to have visited Stabroek again by the route here described. The plan was to have ascended the Amazons from Para and got into the Rio Negro, and from thence to have returned towards the source of the Essequibo, in order to examine the crystal mountains, and look once more for Lake Parima, or the White Sea; but on arriving at Cayenne, the current was running with such amazing rapidity to leeward, that a Portuguese sloop, which had been beating up towards Para for four weeks, was then only half-way. Finding, therefore, that a beat to the Amazons would be long, tedious, and even uncertain, and aware that the season for procuring birds with fine plumage had already set in, I left Cayenne in an American ship for Paramaribo, went through the interior to the Coryntin, stopped a few days in New Amsterdam, and proceeded to Demerara. If, gentle reader, thy patience be not already worn out, and thy eyes half closed in slumber, by perusing the dull adventures of this second sally, perhaps thou wilt pardon a line or two on Demerara; and then we will retire to its forests, to collect and examine the economy of its most rare and beautiful birds, and give the world a new mode of preserving them.

Stabroek, the capital of Demerara, has been rapidly increasing for some years back; and if prosperity go hand in hand with the present enterprising spirit, Stabroek, ere long, will be of the first colonial consideration. It stands on the eastern bank at the mouth of the Demerara, and enjoys all the advantages of the refreshing sea-breeze; the streets are spacious, well-bricked, and elevated, the trenches clean, the bridges excellent, and the houses handsome. Almost every commodity and luxury of London may be bought in the shops at Stabroek; its market wants better regulations. The hotels are commodious, clean, and well attended. Demerara boasts as fine and well-disciplined militia as any colony in the western world.

The court of justice, where, in times of old, the bandage was easily removed from the eyes of the goddess, and her scales thrown out of equilibrium, now rises in dignity under the firmness, talents, and urbanity of Mr. President Rough.

The plantations have an appearance of high cultivation; a tolerable idea may be formed of their value when you know that last year Demerara numbered seventy-two thousand nine hundred and ninety-nine slaves. They made about forty-four million pounds of sugar, near two million gallons of rum, above eleven million pounds of coffee, and three million eight hundred and nineteen thousand five hundred and twelve pounds of cotton; the receipt into the public chest was five hundred and fifty-three thousand nine hundred and fifty-six guilders; the public expenditure, four hundred and fifty-one thousand six hundred and three guilders.

Slavery can never be defended; he whose heart is not of iron can never wish to be able to defend it; while he heaves a sigh for the poor negro in captivity, he wishes from his soul that the traffic had been stifled in its birth; but, unfortunately, the Governments of Europe nourished it, and now that they are exerting themselves to do away the evil, and ensure liberty to the sons of Africa, the situation of the plantation slaves is depicted as truly deplorable, and their condition wretched. It is not so. A Briton's heart, proverbially kind and generous, is not changed by climate, or its streams of compassion dried up by the scorching heat of a Demerara sun; he cheers his negroes in labour, comforts them in sickness, is kind to them in old age, and never forgets that they are his fellow-creatures.

Instances of cruelty and depravity certainly occur here as well as all the world over: but the edicts of the colonial Government are well calculated to prevent them; and the British planter, except here and there one, feels for the wrongs done to a poor ill-treated slave, and shows that his heart grieves for him by causing immediate redress, and preventing a repetition.

Long may ye flourish, peaceful and liberal inhabitants of Demerara! Your doors are ever open to harbour the harbourless; your purses never shut to the wants of the distressed: many a ruined fugitive from the Oroonoque will bless your kindness to him in the hour of need, when, flying from the woes of civil discord, without food or raiment, he begged for shelter underneath your roof. The poor sufferer in Trinidad, who lost his all in the devouring flames, will remember your charity to his latest moments. The traveller, as he leaves your port, casts a longing lingering look behind; your attentions, your hospitality, your pleasantry and mirth, are uppermost in his thoughts; your prosperity is close to his heart. Let us now, gentle reader, retire from the busy scenes of man, and journey on towards the wilds in quest of the feathered tribe.

Leave behind you your high-seasoned dishes, your wines and your delicacies; carry nothing but what is necessary for your own comfort and the object in view, and depend upon the skill of an Indian, or your own, for fish and game. A sheet, about twelve feet long, ten wide, painted, and with loop-holes on each side, will be of great service; in a few minutes you can suspend it betwixt two trees in the shape of a roof. Under this, in your hammock, you may defy the pelting shower, and sleep heedless of the dews of night. A hat, a shirt, and a light pair of trousers, will be all the raiment you require. Custom will soon teach you to tread lightly and barefoot on the little inequalities of ground, and show you how to pass on, unwounded, amid the mantling briers.

Snakes in these wilds are certainly an annoyance, though perhaps more in imagination than reality: for you must recollect, that the serpent is never the first to offend; his poisonous fang was not given him for conquest: he never inflicts a wound with it but to defend existence. Provided you walk cautiously, and do not absolutely touch him, you may pass in safety close by him. As he is often coiled up on the ground, and amongst the branches of the trees above you, a degree of circumspection is necessary, lest you unwarily disturb him.

Tigers are too few, and too apt to fly before the noble face of man, to require a moment of your attention.

The bite of the most noxious of the insects, at the very worst, only causes a transient fever, with a degree of pain more or less.

Birds in general, with few exceptions, are not common in the very remote parts of the forest. The sides of rivers, lakes, and creeks, the borders of savannas, the old abandoned habitations of Indians and woodcutters, seem to be their favourite haunts.

Though least in size, the glittering mantle of the humming-bird entitles it to the first place in the list of the birds of the New World. It may truly be called the Bird of Paradise; and had it existed in the Old World, it would have claimed the title instead of the bird which has now the honour to bear it. See it darting through the air almost as quick as thought!—now it is within a yard of your face!—in an instant gone!—now it flutters from flower to flower to sip the silver dew—it is now a ruby—now a topaz—now an emerald—now all burnished gold! It would be arrogant to pretend to describe this winged gem of nature after Buffon's elegant description of it.

Cayenne and Demerara produce the same humming-birds. Perhaps you would wish to know something of their haunts. Chiefly in the months of July and August the tree called *bois immortel*, very common in Demerara, bears abundance of red blossom, which stays on the tree some weeks; then it is that most of the different species of humming-birds are very plentiful. The wild red sage is also their favourite shrub, and they buzz like bees around the blossom of the wallaba-tree. Indeed, there is scarce a flower in the interior, or on the sea-coast, but what receives frequent visits from one or other of the species.

On entering the forests, on the rising land in the interior, the blue and green, the smallest brown, no bigger than the humblebee, with two long feathers in the tail, and the little forked-tail purple-throated humming-birds, glitter before you in ever-changing attitudes. One species alone never shows his beauty to the sun; and were it not for his lovely shining colours, you might almost be tempted to class him with the goat suckers on account of his habits. He is the largest of all the humming-birds, and is all red and changing gold-green, except the head, which is black. He has two long feathers in the tail, which cross each other, and these have gained him the name of karabimiti, or ara humming-bird, from the Indians. You will never find him on the sea-coast, or where the river is salt, or in the heart of the forest, unless fresh water be there. He keeps close by the side of woody fresh-water rivers and dark and lonely creeks. He leaves his retreat before sunrise to feed on the insects over the water; he returns to it as soon as the sun's rays cause a glare of light, is sedentary all day long, and comes out again for a short time after sunset. He builds his nest on a twig over the water in the unfrequented creeks; it looks like tanned cow-leather.

As you advance towards the mountains of Demerara, other species of humming-birds present themselves before you. It seems to be an erroneous opinion that the humming-bird lives entirely on honey-dew. Almost every flower of the tropical climate contains insects of one kind or other; now, the humming-bird is most busy about the flowers an hour or two after sunrise and after a shower of rain, and it is just at this time that the insects come out to the edge of the flower in order that the sun's rays may dry the nocturnal dew and rain which they have received. On opening the stomach of the humming-bird, dead insects are almost always found there.

Next to the humming-birds, the cotingas display the gayest plumage. They are of the order of passeres, and you number five species betwixt the sea-coast and the rock Saba. Perhaps the scarlet cotinga is the richest of the five, and is one of those birds which are found in the deepest recesses of the forest. His crown is flaming red; to this abruptly succeeds a dark shining brown, reaching half-way down the back: the remainder of the back, the rump, and tail, the extremity of which is edged with black, are a lively red; the belly is a somewhat lighter red; the breast, reddish-black; the wings, brown. He has no song, is solitary, and utters a monotonous whistle which sounds like "quet." He is fond of the seeds of the hitia-tree, and those of the siloabali and bastard-siloabali trees, which ripen in December, and continue on the trees for about two months. He is found throughout the year in Demerara; still nothing is known of his incubation. The Indians all agree in telling you that they have never seen his nest.

The purple-breasted cotinga has the throat and breast of a deep purple, the wings and tail black, and all the rest of the body a most lively shining blue.

The purple-throated cotinga has black wings and tail, and every other part a light and glossy blue, save the throat, which is purple.

The pompadour cotinga is entirely purple, except his wings, which are white, their four first feathers tipped with brown. The great coverts of the wings are stiff, narrow, and pointed, being shaped quite different from those of any other bird. When you are betwixt this bird and the sun in his flight, he appears uncommonly brilliant. He makes a hoarse noise, which sounds like "wallababa." Hence his name amongst the Indians.

None of these three cotingas have a song. They feed on the hitia, siloabali, and bastard-siloabali seeds, the wild guava, the fig, and other fruit trees of the forest. They are easily shot in these trees during the months of December, January, and part of February. The greater part of them disappear after this, and probably retire far away to breed. Their nests have never been found in Demerara.

The fifth species is the celebrated campanero of the Spaniards, called dara by the Indians and bell-bird by the English. He is about the size of the jay. His plumage is white as snow. On his forehead rises a spiral tube nearly three inches long. It is jet-black, dotted all over with small white feathers. It has a communication with the palate, and when filled with air, looks like a spire; when empty, it becomes pendulous. His note is loud and clear, like the sound of a bell, and may be heard at the distance of three miles. In the midst of these extensive wilds, generally on the dried top of an aged mora, almost out of gun reach, you will see the campanero. No sound or song from any of the winged inhabitants of the forest, not even the clearly-pronounced "Whip-poor-Will," from the goatsucker, causes such astonishment as the toll of the campanero.

With many of the feathered race, he pays the common tribute of a morning and an evening song; and even when the meridian sun has shut in silence the mouths of almost the whole of animated nature, the campanero still cheers the forest. You hear his toll, and then a pause for a minute, then another toll, and then a pause again, and then a toll, and again a pause. Then he is silent for six or eight minutes, and then, another toll, and so on. Actæon would stop in mid chase, Maria would defer her evening song, and

Orpheus himself would drop his lute to listen to him, so sweet, so novel, and romantic is the toll of the pretty snow-white campanero. He is never seen to feed with the other cotingas, nor is it known in what part of Guiana he makes his nest.

While cotingas attract your attention by their superior plumage, the singular form of the toucan makes a lasting impression on your memory. There are three species of toucans in Demerara, and three diminutives, which may be called toucanets. The largest of the first species frequents the mangrove-trees on the sea-coast. He is never seen in the interior till you reach Macoushia, where he is found in the neighbourhood of the river Tacatou. The other two species are very common. They feed entirely on the fruits of the forest, and though of the pie kind, never kill the young of other birds or touch carrion. The larger is called bouradi by the Indians (which means "nose"), the other, scirou. They seem partial to each other's company, and often resort to the same feeding tree, and retire together to the same shady noon-day retreat. They are very noisy in rainy weather at all hours of the day, and in fair weather, at morn and eve. The sound which the bouradi makes is like the clear yelping of a puppy dog, and you fancy he says, "pia-po-o-co," and thus the South American Spaniards call him piapoco.

All the toucanets feed on the same trees on which the toucan feeds, and every species of this family of enormous bill lays its eggs in the hollow trees. They are social, but not gregarious. You may sometimes see eight or ten in company, and from this you would suppose they are gregarious; but, upon a closer examination, yon will find it has only been a dinner party, which breaks up and disperses towards roosting-time.

You will be at a loss to conjecture for what ends nature has overloaded the head of this bird with such an enormous bill. It cannot be for the offensive, as it has no need to wage war with any of the tribes of animated nature; for its food is fruits and seeds, and those are in superabundance throughout the whole year in the regions where the toucan is found. It can hardly be for the defensive, as the toucan is preyed upon by no bird in South America, and were it obliged to be at war, the texture of the bill is ill adapted to give or receive blows, as you will see in dissecting it. It cannot be for any particular protection to the tongue, as the tongue is a perfect feather.

The flight of the toucan is by jerks; in the action of flying it seems incommoded by this huge disproportioned feature, and the head seems as if bowed down to the earth by it against its will; if the extraordinary form and size of the bill expose the toucan to ridicule, its colours make it amends. Were a specimen of each species of the toucan presented to you, you would pronounce the bill of the bouradi the most rich and beautiful; on the ridge of the upper mandible a broad stripe of most lovely yellow extends from the head to the point; a stripe of the same breadth, though somewhat deeper yellow, falls from it at right angles next the head down to the edge of the mandible; then follows a black stripe, half as broad, falling at right angles from the ridge, and running narrower along the edge to within half an inch of the point. The rest of the mandible is a deep bright red. The lower mandible has no yellow; its black and red are distributed in the same manner as on the upper one, with this difference, that there is black about an inch from the point. The stripe corresponding to the deep yellow stripe on the upper mandible is sky blue. It is worthy of remark that all these brilliant colours of the bill are to be found in the plumage of the body, and the bare skin round the eye.

All these colours, except the blue, are inherent in the horn; that part which appears blue is in reality transparent white, and receives its colour from a thin piece of blue skin inside. This superb bill fades in death, and in three or four days' time has quite lost its original colours.

Till within these few years no idea of the true colours of the bill could be formed from the stuffed toucans brought to Europe. About eight years ago, while eating a boiled toucan, the thought struck me that the colours in the bill of a preserved specimen might be kept as bright as those in life. A series of experiments proved this beyond a doubt. If you take your penknife and cut away the roof of the upper mandible, you will find that the space betwixt it and the outer shell contains a large collection of veins, and small osseous fibres running in all directions through the whole extent of the bill. Clear away all these with your knife, and you will come to a substance more firm than skin, but of not so strong a texture as the horn itself; cut this away also, and behind it is discovered a thin and tender membrane: yellow where it has touched the yellow part of the horn, blue where it has touched the red part, and black towards the edge and point. When dried, this thin and tender membrane becomes nearly black; as soon as it is cut away, nothing remains but the outer horn, red and yellow, and now become transparent. The under mandible must undergo the same operation. Great care must be taken, and the knife used very cautiously when you are cutting through the different parts close to where the bill joins on to the head: if you cut away too much, the bill drops off; if you press too hard, the knife comes through the horn; if you leave too great a portion of the membrane, it appears through the horn, and by becoming black when dried, makes the horn appear black also, and has a bad effect. Judgment, caution, skill, and practice will ensure success.

You have now cleared the bill of all those bodies which are the cause of its apparent fading; for, as has been said before, these bodies dry in death, and become quite discoloured, and appear so through the horn; and reviewing the bill in this state, you conclude that its former bright colours are lost.

Something still remains to be done. You have rendered the bill transparent by the operation, and that transparency must be done away to make it appear perfectly natural. Pound some clean chalk, and give it enough water till it be of the consistency of tar; add a proportion of gum-arabic to make it adhesive; then take a camel-hair brush, and give the inside of both mandibles a coat; apply a second when the first is dry, then another, and a fourth to finish all. The gum-arabic will prevent the chalk from cracking and falling off. If you remember, there is a little space of transparent white on the lower mandible which originally appeared blue, but which became transparent white as soon as the thin piece of blue skin was cut away; this must be painted blue inside. When all this is completed, the bill will please you; it will appear in its original colours. Probably your own abilities will suggest a cleverer mode of operating than the one here described. A small gouge would assist the penknife, and render the operation less difficult.

The houtou ranks high in beauty amongst the birds of Demerara—his whole body is green, with a bluish cast in the wings and tail; his crown, which he erects at pleasure, consists of black in the centre, surrounded with lovely blue of two different shades: he has a triangular black spot, edged with blue, behind the eye, extending to the ear; and on his breast a sable tuft, consisting of nine feathers edged also with blue. This bird seems to suppose that its beauty can be increased by trimming the tail, which undergoes the same operation as our hair in a barber's shop, only with this difference, that it uses its own beak, which is serrated, in lieu of a pair of scissors; as soon as his tail is full grown, he begins about an inch from the extremity of the two longest feathers in it, and cuts away the web on both sides of the shaft, making a gap about an inch long; both male and female Adonise their tails in this manner, which gives them a remarkable appearance amongst all other birds. While we consider the tail of the houtou blemished and defective, were he to come amongst us he would probably consider our heads, cropped and bald, in no better light. He who wishes to observe this handsome bird in his native haunts must be in the forest at the morning's dawn. The houtou shuns the society of man: the plantations and cultivated parts are too much disturbed to engage it to settle there; the thick and gloomy forests are the places preferred by the solitary houtou. In those far-extending wilds, about daybreak, you hear him articulate, in a distinct and mournful tone, "Houtou, houtou." Move cautiously on to where the sound proceeds from, and you will see him sitting in the underwood, about a couple of yards from the ground, his tail moving up and down every time he articulates "houtou." He lives on insects and the berries amongst the underwood, and very rarely is seen in the lofty trees, except the bastard-siloabali tree, the fruit of which is grateful to him. He makes no nest, but rears his young in a hole in the sand, generally on the side of a hill.

While in quest of the houtou you will now and then fall in with the jay of Guiana, called by the Indians ibibirou. Its forehead is black, the rest of the head white; the throat and breast like the English magpie: about an inch of the extremity of the tail is white, the other part of it, together with the back and wings, a greyish changing purple; the belly is white: there are generally six or eight of them in company; they are shy and garrulous, and tarry a very short time in one place; they are never seen in the cultivated parts.

Through the whole extent of the forest, chiefly from sunrise till nine o'clock in the morning, you hear a sound of "Wow, wow, wow, wow." This is the bird called boclora by the Indians. It is smaller than the common pigeon, and seems, in some measure, to partake of its nature; its head and breast are blue; the back and rump somewhat resemble the colour on the peacock's neck; its belly is a bright yellow; the legs are so very short that it always appears as if sitting on the branch; it is as ill-adapted for walking as the swallow; its neck, for above an inch all round, is quite bare of feathers, but this deficiency is not seen, for it always sits with its head drawn in upon its shoulders: it sometimes feeds with the cotingas on the guava and hitia trees; but its chief nutriment seems to be insects, and, like most birds which follow this prey, its chaps are well armed with bristles: it is found in Demerara at all times of the year, and makes a nest resembling that of the stock-dove. This bird never takes long flights, and when it crosses a river or creek it goes by long jerks.

The boclora is very unsuspicious, appearing quite heedless of danger: the report of a gun within twenty yards will not cause it to leave the branch on which it is sitting, and you may often approach it so near as almost to touch it with the end of your bow. Perhaps there is no bird known whose feathers are so slightly fixed to the skin as those of the boclora. After shooting it, if it touch a branch in its descent, or if it drop on hard ground, whole heaps of feathers fall off; on this account it is extremely hard to procure a specimen for preservation. As soon as the skin is dry in the preserved specimen, the feathers become as well fixed as those in any other bird.

Another species, larger than the boclora, attracts much of your notice in these wilds; it is called cuia by the Indians, from the sound of its voice; its habits are the same as those of the boclora, but its colours different; its head, breast, back, and rump are a shining, changing green; its tail not quite so bright; a black bar runs across the tail towards the extremity; and the outside feathers are partly white, as in the boclora; its belly is entirely vermilion, a bar of white separating it from the green on the breast.

There are diminutives of both these birds; they have the same habits, with a somewhat different plumage, and about half the size. Arrayed from head to tail in a robe of richest sable hue, the bird called rice-bird loves spots cultivated by the hand of man. The woodcutter's house on the hills in the interior, and the planter's habitation on the sea-coast, equally attract this songless species of the order of pie, provided the Indian corn be ripe there. He is nearly of the jackdaw's size, and makes his nest far away from the haunts of men; he may truly be called a blackbird: independent of his plumage, his beak, inside and out, his legs, his toes, and claws, are jet black.

Mankind, by clearing the ground, and sowing a variety of seeds, induces many kinds of birds to leave their native haunts and come and settle near him; their little depredations on his seeds and fruits prove that it is the property, and not the proprietor, which has the attractions.

One bird, however, in Demerara, is not actuated by selfish motives; this is the cassique; in size, he is larger than the starling; he courts the society of man, but disdains to live by his labours. When nature calls for support, he repairs to the neighbouring forest, and there partakes of the store of fruits and seeds which she has produced in abundance for her aërial tribes. When his repast is over, he returns to man, and pays the little tribute which he owes him for his protection; he takes his station on a tree close to his house, and there, for hours together, pours forth a succession of imitative notes. His own song is sweet, but very short. If a toucan be yelping in the neighbourhood, he drops it, and imitates him. Then he will amuse his protector with the cries of the different species of the woodpecker; and when the sheep bleat, he will distinctly answer them. Then comes his own song again; and if a puppy-dog or a Guinea-fowl interrupt him, he takes them off admirably, and by his different gestures during the time, you would conclude that he enjoys the sport.

The cassique is gregarious, and imitates any sound he hears with such exactness, that he goes by no other name than that of mocking-bird amongst the colonists.

At breeding-time, a number of these pretty choristers resort to a tree near the planter's house, and from its outside branches weave their pendulous nests. So conscious do they seem that they never give offence, and so little suspicious are they of receiving any injury from man, that they will choose a tree within forty yards from his house, and occupy the branches so low down, that he may peep into the nests. A tree in Waratilla creek affords a proof of this.

The proportions of the cassique are so fine, that he may be said to be a model of symmetry in ornithology. On each wing he has a bright yellow spot, and his rump, belly, and half the tail are of the same colour. All the rest of the body is black. His beak is the colour of sulphur, but it fades in death, and requires the same operation as the bill of the toucan to make it keep its colours. Up the rivers, in the interior, there is another cassique, nearly the same size, and of the same habits, though not gifted with its powers of imitation. Except in breeding time, you will see hundreds of them retiring to roost, amongst the mocamoca-trees and low shrubs on the banks of the Demerara, after you pass the first island. They are not common on the sea-coast. The rump of the cassique is a flaming scarlet. All the rest of the body is a rich glossy black. His bill is sulphur colour. You may often see numbers of this species weaving their pendulous nests on one side of a tree, while numbers of the other species are busy in forming theirs on the opposite side of the same tree. Though such near neighbours, the females are never observed to kick up a row, or come to blows!

Another species of cassique, as large as a crow, is very common in the plantations. In the morning he generally repairs to a large tree, and there, with his tail spread over his back, and shaking his lowered wings, he produces notes which, though they cannot be said to amount to a song, still have something very sweet and pleasing in them. He makes his nest in the same form as the other cassiques. It is above four feet long; and when you pass under the tree, which often contains fifty or sixty of them, you cannot help stopping to admire them as they wave to and fro, the sport of every storm and breeze. The rump is chestnut; ten feathers of the tail are a fine yellow, the remaining two, which are the middle ones, are black, and an inch shorter than the others. His bill is sulphur colour; all the rest of the body black, with here and there shades of brown. He has five or six long narrow black feathers on the back of his head, which he erects at pleasure.

There is one more species of cassique in Demerara, which always prefers the forest to the cultivated parts. His economy is the same as that of the other cassiques. He is rather smaller than the last described bird. His body is greenish, and his tail and rump paler than those of the former. Half of his beak is red.

You would not be long in the forests of Demerara without noticing the woodpeckers. You meet with them feeding at all hours of the day. Well may they do so. Were they to follow the example of most of the other birds, and only feed in the morning and evening, they would be often on short allowance, for they sometimes have to labour three or four hours at the tree before they get to their food. The sound which the largest kind makes in hammering against the bark of the tree is so loud, that you would never suppose it to proceed from the efforts of a bird. You would take it to be the woodman, with his axe, trying by a sturdy blow, often repeated, whether the tree were sound or not. There are fourteen species here; the largest the size of a magpie, the smallest no bigger than the wren. They are all beautiful; and the greater part of them have their heads ornamented with a fine crest, movable at pleasure.

It is said, if you once give a dog a bad name, whether innocent or guilty, he never loses it; it sticks close to him wherever he goes. He has many a kick and many a blow to bear on account of it; and there is nobody to stand up for him. The woodpecker is little better off. The proprietors of woods in Europe have long accused him of injuring their timber, by boring holes in it, and letting in the water, which soon rots it. The colonists in America have the same complaint against him. Had he the power of speech, which Ovid's birds possessed in days of yore, he could soon make a defence. "Mighty lord of the woods," he would say to man, "why do you wrongfully accuse me? why do you hunt me up and down to death for an imaginary offence? I have never spoiled a leaf of your property, much less your wood. Your merciless shot strikes me at the very time I am doing you a service. But your short-sightedness will not let you see it, or your pride is above examining closely the actions of so insignificant a little bird as I am. If there be that spark of feeling in your breast which they say man possesses, or ought to possess, above all other animals, do a poor injured creature a little kindness, and watch me in your woods only for one day. I never wound your healthy trees. I should perish for want in the attempt. The sound bark would easily resist the force of my bill: and were I even to pierce through it, there would be nothing inside that I could fancy, or my stomach digest. I often visit them, it is true, but a knock or two convince me that I must go elsewhere for support; and were you to listen attentively to the sound which my bill causes, you would know whether I am upon a healthy or an unhealthy tree. Wood and bark are not my food. I live entirely upon the insects which have already formed a lodgment in the distempered tree. When the sound informs me that my prey is there, I labour for hours together till I get at it; and by consuming it, for my own support, I prevent its further depredations in that part. Thus I discover for you your hidden and unsuspected foe, which has been devouring your wood in such secrecy, that you had not the least suspicion it was there. The hole which I make in order to get at the pernicious vermin will be seen by you as you pass under the tree. I leave it as a signal to tell you that your tree has already stood too long. It is past its prime. Millions of insects, engendered by disease, are preying upon its vitals. Ere long it will fall a log in useless ruins. Warned by this loss, cut down the rest in time, and spare, O spare the unoffending woodpecker."

In the rivers, and different creeks, you number six species of the kingfisher. They make their nest in a hole in the sand on the side of the bank. As there is always plenty of foliage to protect them from the heat of the sun, they feed at all hours of the day. Though their plumage is prettily varied, still it falls far short of the brilliancy displayed by the English kingfisher. This little native of Britain would outweigh them altogether in the scale of beauty.

A bird called jacamar is often taken for a kingfisher, but it has no relationship to that tribe; it frequently sits in the trees over the water, and as its beak bears some resemblance to that of the kingfisher, this may probably account for its being taken for one; it feeds entirely upon insects; it sits on a branch in motionless expectation, and as soon as a fly, butterfly, or moth passes by, it darts at it, and returns to the branch it had just left. It seems an indolent, sedentary bird, shunning the society of all others in the forest. It never visits the plantations, but is found at all times of the year in the woods. There are four species of jacamar in Demerara; they are all beautiful; the largest, rich and superb in the extreme. Its plumage is of so fine a changing blue and golden green, that it may be ranked with the choicest of the humming birds. Nature has denied it a song, but given a costly garment in lieu of it. The smallest species of jacamar is very common in the dry savannas. The second size, all golden green on the back, must be looked for in the wallaba forest. The third is found throughout the whole extent of these wilds; and the fourth, which is the largest, frequents the interior, where you begin to perceive stones in the ground.

When you have penetrated far into Macoushia, you hear the pretty songster called troupiale pour forth a variety of sweet and plaintive notes. This is the bird which the Portuguese call the nightingale of Guiana; its predominant colours are rich orange and shining black, arrayed to great advantage; his delicate and well-shaped frame seems unable to bear captivity. The Indians sometimes bring down troupiales to Stabroek, but in a few months they languish and die in a cage. They soon become very familiar; and if you allow them the liberty of the house, they live longer than in a cage, and appear in better spirits; but, when you least expect it, they drop down and die in epilepsy.

Smaller in size, and of colour not so rich and somewhat differently arranged, another species of troupiale sings melodiously in Demerara. The woodcutter is particularly favoured by him; for while the hen is sitting on her nest, built in the roof of the woodcutter's house, he sings for hours together close by: he prefers the forests to the cultivated parts.

You would not grudge to stop for a few minutes, as you are walking in the plantations, to observe a third species of troupiale: his wings, tail, and throat are black; all the rest of the body is a bright yellow. There is something very sweet and plaintive in his song, though much shorter than that of the troupiale in the interior.

A fourth species goes in flocks from place to place in the cultivated parts at the time the Indian corn is ripe; he is all black, except the head and throat, which are yellow; his attempt at song is not worth attending to.

Wherever there is a wild fig-tree ripe, a numerous species of birds, called tangara, is sure to be on it. There are eighteen beautiful species here. Their plumage is very rich and diversified; some of them boast six separate colours; others have the blue, purple, green, and black so kindly blended into each other, that it would be impossible to mark their boundaries; while others again exhibit them strong, distinct, and abrupt: many of these tangaras have a fine song. They seem to partake much of the nature of our linnets, sparrows, and finches. Some of them are fond of the plantations; others are never seen there, preferring the wild seeds of the forest to the choicest fruits planted by the hand of man.

On the same fig-trees to which they repair, and often accidentally up and down the forest, you fall in with four species of manikin. The largest is white and black, with the feathers on the throat remarkably long; the next in size is half red and half black; the third, black, with a white crown; the fourth, black, with a golden crown, and red feathers at the knee. The half red and half black species is the scarcest. There is a creek in the Demerara called Camouni. About ten minutes from the mouth, you see a common-sized fig-tree on your right-hand, as you ascend, hanging over water; it bears a very small fig twice a year. When its fruit is ripe, this manikin is on the tree from morn till eve.

On all the ripe fig-trees in the forest you see the bird called the small tiger-bird. Like some of our belles and dandies, it has a gaudy vest to veil an ill-shaped body: the throat, and part of the head, are a bright red; the breast and belly have black spots on a yellow ground; the wings are a dark green, black, and white; and the rump and tail black and green. Like the manikin, it has no song: it depends solely upon a showy garment for admiration.

Devoid, too, of song, and in a still superber garb, the yawaraciri comes to feed on the same tree. It has a bar like black velvet from the eyes to the beak; its legs are yellow; its throat, wings, and tail black; all the rest of the body a charming blue. Chiefly in the dry savannas, and here and there accidentally in the forest, you see a songless yawaraciri still lovelier than the last: his crown is whitish blue, arrayed like a coat of mail; his tail is black, his wings black and yellow, legs red, and the whole body a glossy blue.

Whilst roving through the forest, ever and anon you see individuals of the wren species busy amongst the fallen leaves, or seeking insects at the roots of the trees.

Here, too, you find six or seven species of small birds, whose backs appear to be overloaded with silky plumage. One of these, with a chestnut breast, smoke-coloured back, tail red, white feathers like horns on his head, and white narrow-pointed feathers under the jaw, feeds entirely upon ants. When a nest of large, light brown ants emigrates, one following the other in meandering lines above a mile long, you see this bird watching them, and every now and then picking them up. When they disappear he is seen no more: perhaps this is the only kind of ant he is fond of; when these ants are stirring, you are sure to find him near them. You cannot well mistake the ant after you have once been in its company, for its sting is very severe, and you can hardly shoot the bird, and pick it up, without having five or six upon you.

Parrots and paroquets are very numerous here, and of many different kinds. You will know when they are near you in the forest, not only by the noise they make, but also by the fruits and seeds which they let fall while they are feeding.

The hia-hia parrot, called in England the "parrot of the sun," is very remarkable: he can erect at pleasure a fine radiated circle of tartan feathers quite round the back of his head from jaw to jaw. The fore-part of his head is white: his back, tail, and wings green; and his breast and belly tartan.

Superior in size and beauty to every parrot of South America, the ara will force you to take your eyes from the rest of animated nature and gaze at him: his commanding strength, the flaming scarlet of his body, the lovely variety of red, yellow, blue, and green in his wings, the extraordinary length of his scarlet and blue tail seem all to join and demand for him the title of "emperor of all the parrots." He is scarce in Demerara till you reach the confines of the Macoushi country; there he is in vast abundance; he mostly feeds on trees of the palm species. When the coucourite-trees have ripe fruit on them, they are covered with this magnificent parrot: he is not shy or wary; you may take your blow-pipe and quiver of poisoned arrows, and kill more than you are able to carry back to your hut. They are very vociferous, and, like the common parrots, rise up in bodies towards sunset, and fly two and two to their place of rest. It is a grand sight in ornithology to see thousands of aras flying over your head, low enough to let you have a full view of their flaming mantle. The Indians find their flesh very good, and the feathers serve for ornaments in their head-dresses. They breed in the holes of trees, are easily reared and tamed, and learn to speak pretty distinctly.

Another species frequents the low lands of Demerara. He is nearly the size of the scarlet ara, but much inferior in plumage. Blue and yellow are his predominant colours.

Along the creeks and river sides, and in the wet savannas, six species of the bittern will engage your attention. They are all handsome. The smallest is not so large as the English water-hen.

In the savannas, too, you will sometimes surprise the snow-white egrette, whose back is adorned with the plumes from which it takes its name. Here, too, the spur-winged water-hen, the blue and green water-hen, and two other species of ordinary plumage are found. While in quest of these, the blue heron, the large and small brown heron, the boat-bill, and Muscovy duck now and then rise up before you.

When the sun has sunk in the western woods, no longer agitated by the breeze; when you can only see a straggler or two of the feathered tribe hastening to join its mate, already at its roosting-place, then it is that the goatsucker comes out of the forest, where it has sat all day long in slumbering ease, unmindful of the gay and busy scenes around it. Its eyes are too delicately formed to bear the light, and thus it is forced to shun the flaming face of day, and wait in patience till night invites him to partake of the pleasures her dusky presence brings.

The harmless, unoffending goatsucker, from the time of Aristotle down to the present day, has been in disgrace with man. Father has handed down to son, and author to author, that this nocturnal thief subsists by milking the flocks. Poor injured little bird of night, how sadly hast thou suffered, and how foul a stain has inattention to facts put upon thy character! Thou hast never robbed man of any part of his property, nor deprived the kid of a drop of milk.

When the moon shines bright you may have a fair opportunity of examining the goatsucker. You will see it close by the cows, goats, and sheep, jumping up every now and then under their bellies. Approach a little nearer—he is not shy, "he fears no danger, for he knows no sin." See how the nocturnal flies are tormenting the herd, and with what dexterity he springs up and catches them, as fast as they alight on the belly, legs, and udder of the animals. Observe how quiet they stand, and how sensible they seem of his good offices, for they neither strike at him, nor hit him with their tail, nor tread on him, nor try to drive him away as an uncivil intruder. Were you to dissect him, and inspect his stomach, you would find no milk there. It is full of the flies which have been annoying the herd.

The prettily-mottled plumage of the goatsucker, like that of the owl, wants the lustre which is observed in the feathers of the birds of day. This at once marks him as a lover of the pale moon's nightly beams. There are nine species here. The largest appears nearly the size of the English wood-owl. Its cry is so remarkable that, having once heard it, you will never forget it. When night reigns over these immeasurable wilds, whilst lying in your hammock, you will hear this goatsucker lamenting like one in deep distress. A stranger would never conceive it to be the cry of a bird. He would say it was the departing voice of a midnight-murdered victim, or the last wailing of Niobe for her poor children, before she was turned into stone. Suppose yourself in hopeless sorrow, begin with a high loud note, and pronounce, "Ha, ha, ha, ha, ha, ha, ha," each note lower and lower, till the last is scarcely heard, pausing a moment or two betwixt every note, and you will have some idea of the moaning of the largest goatsucker in Demerara.

Four other species of the goatsucker articulate some words so distinctly, that they have received their names from the sentences they utter, and absolutely bewilder the stranger on his arrival in these parts. The most common one sits down close by your door, and flies and alights three or four yards before yon, as you walk along the road, crying, "Who-are-you, who-who-who-are-you?" Another bids you, "Work-away, work—work-work-away." A third cries mournfully, "Willy-come-go. Willy-Willy-Willy-come-go." And high up in the country, a fourth tells you to "Whip-poor-Will. Whip-whip-whip-poor-Will."

You will never persuade the negro to destroy these birds, or get the Indian to let fly his arrow at them. They are birds of omen and reverential dread. Jumbo, the demon of Africa, has them under his command; and they equally obey the Yabahou, or Demerara Indian devil. They are the receptacles for departed souls, who come back again to earth, unable to rest for crimes done in their days of nature; or they are expressly sent by Jumbo, or Yabahou, to haunt cruel and hard-hearted masters, and retaliate injuries received from them. If the largest goatsucker chance to cry near the white man's door, sorrow and grief will soon be inside; and they expect to see the master waste away with a slow consuming sickness. If it be heard close to the negro's or Indian's hut, from that night misfortune sits brooding over it; and they await the event in terrible suspense.

You will forgive the poor Indian of Guiana for this. He knows no better; he has nobody to teach him. But shame it is that, in our own civilised country, the black cat and broomstaff should be considered as conductors to and from the regions of departed spirits.

Many years ago I knew poor harmless Mary; old age had marked her strongly, just as he will mark you and me, should we arrive at her years and carry the weight of grief which bent her double. The old men of the village said she had been very pretty in her youth; and nothing could be seen more comely than Mary when she danced on the green. He who had gained her heart left her for another, less fair, though richer than Mary. From that time she became sad and pensive; the rose left her cheek, and she was never more seen to dance round the May-pole on the green: her expectations were blighted; she became quite indifferent to everything around her, and seemed to think of nothing but how she could best attend her mother, who was lame, and not long for this life. Her mother had begged a black kitten from some boys who were going to drown it, and in her last illness she told Mary to be kind to it for her sake.

When age and want had destroyed the symmetry of Mary's fine form, the village began to consider her as one who had dealings with spirits; her cat confirmed the suspicion. If a cow died, or a villager wasted away with an unknown complaint, Mary and her cat had it to answer for. Her broom sometimes served her for a walking-stick; and if over she supported her tottering frame with it as far as the May-pole, where once, in youthful bloom and beauty, she had attracted the eyes of all, the boys would surround her and make sport of her, while her cat had neither friend nor safety beyond the cottage wall. Nobody considered it cruel or uncharitable to torment a witch; and it is probable, long before this, that cruelty, old age, and want have worn her out, and that both poor Mary and her cat have ceased to be.

Would you wish to pursue the different species of game, well stored and boundless is your range in Demerara. Here no one dogs you, and afterwards clandestinely inquires if you have a hundred a year in land to entitle you to enjoy such patrician sport. Here no saucy intruder asks if you have taken out a licence, by virtue of which you are allowed to kill the birds which have bred upon your own property. Here

> "You are as free as when God first made man,
> Ere the vile laws of servitude began,
> And wild in woods the noble savage ran."

Before the morning's dawn you hear a noise in the forest, which sounds like "duraquaura" often repeated. This is the partridge, a little smaller, and differing somewhat in colour from the English partridge; it lives entirely in the forest, and probably the young brood very soon leave their parents, as you never flush more than two birds in the same place, and in general only one.

About the same hour, and sometimes even at midnight, you hear two species of maam, or tinamou, send forth their long and plaintive whistle from the depth of the forest. The flesh of both is delicious. The largest is plumper, and almost equals in size the black cock of Northumberland. The quail is said to be here, though rare.

The hannaquoi, which some have compared to the pheasant, though with little reason, is very common.

Here are also two species of the powise, or hocco, and two of the small wild turkeys called maroudi; they feed on the ripe fruits of the forest, and are found in all directions in these extensive wilds. You will admire the horned screamer as a stately and majestic bird: he is almost the size of the turkey cock; on his head is a long slender horn, and each wing is armed with a strong, sharp, triangular spur, an inch long.

Sometimes you will fall in with flocks of two or three hundred waracabas, or trumpeters, called so from the singular noise they produce. Their breast is adorned with beautiful changing blue and purple feathers; their head and neck like velvet; their wings and back grey, and belly black. They run with great swiftness, and, when domesticated, attend their master in his walks with as much apparent affection as his dog. They have no spurs, but still, such is their high spirit and activity, that they browbeat every dunghill fowl in the yard, and force the Guinea birds, dogs, and turkeys to own their superiority.

If, kind and gentle reader, thou shouldst ever visit these regions with an intention to examine their productions, perhaps the few observations contained in these Wanderings may be of service to thee; excuse their brevity; more could have been written, and each bird more particularly described, but it would have been pressing too hard upon thy time and patience.

Soon after arriving in these parts, thou wilt find that the species here enumerated are only as a handful from a well-stored granary. Nothing has been said of the eagles, the falcons, the hawks, and shrikes; nothing of the different species of vultures, the king of which is very handsome, and seems to be the only bird which claims regal honours from a surrounding tribe. It is a fact beyond all dispute, that when the scent of carrion has drawn together hundreds of the common vultures, they all retire from the carcase as soon as the king of the vultures makes his appearance. When his majesty has satisfied the cravings of his royal stomach with the choicest bits from the most stinking and corrupted parts, he generally retires to a neighbouring tree, and then the common vultures return in crowds to gobble down his leavings. The Indians, as well as the whites, have observed this, for when one of them, who has learned a little English, sees the king, and wishes you to have a proper notion of the bird, he says, "There is the governor of the carrion crows."

Now, the Indians have never heard of a personage in Demerara higher than that of governor; and the colonists, through a common mistake, call the vultures carrion crows. Hence the Indian, in order to express the dominion of this bird over the common vultures, tells you he is governor of the carrion crows. The Spaniards have also observed it; for, through all the Spanish Main, he is called "rey de zamuros"—king of the vultures. The many species of owls, too, have not been noticed; and no mention made of the Columbine tribe. The prodigious variety of waterfowl on the sea-shore has been but barely hinted at.

There, and on the borders and surface of the inland waters, in the marshes and creeks, besides the flamingos, scarlet curlew, and spoonbills, already mentioned, will be found greenish-brown curlews, sandpipers, rails, coots, gulls, pelicans, jabirus, nandapoas, crabiers, snipes, plovers, ducks, geese, cranes, and anhingas, most of them in vast abundance; some frequenting only the sea-coast, others only the interior, according to their different natures; all worthy the attention of the naturalist, all worthy of a place in the cabinet of the curious.

Should thy comprehensive genius not confine itself to birds alone, grand is the appearance of other objects all around. Thou art in a land rich in botany and mineralogy, rich in zoology and entomology. Animation will glow in thy looks, and exercise will brace thy frame in vigour. The very time of thy absence from the tables of heterogeneous luxury will be profitable to thy stomach, perhaps already sorely drenched with Londo-Parisian sauces, and a new stock of health will bring thee an appetite to relish the wholesome food of the chase; never-failing sleep will wait on thee at the time she comes to soothe the rest of animated nature; and, ere the sun's rays appear in the horizon, thou wilt spring from thy hammock fresh as April lark. Be convinced, also, that the dangers and difficulties which are generally supposed to accompany the traveller in his journey through distant regions, are not half so numerous or dreadful as they are commonly thought to be.

The youth who incautiously reels into the lobby of Drury Lane, after leaving the table sacred to the god of wine, is exposed to more certain ruin, sickness, and decay than he who wanders a whole year in the wilds of Demerara. But this will never be believed; because the disasters arising from dissipation are so common and frequent in civilised life, that man becomes quite habituated to them; and sees daily victims sink into the tomb long before their time, without ever once taking alarm at the causes which precipitated them headlong into it.

But the dangers which a traveller exposes himself to in foreign parts are novel, out-of-the-way things to a man at home. The remotest apprehension of meeting a tremendous tiger, of being carried off by a flying dragon, or having his bones picked by a famished cannibal—oh, that makes him shudder. It sounds in his ears like the bursting of a bomb-shell. Thank Heaven, he is safe by his own fireside!

Prudence and resolution ought to be the traveller's constant companions. The first will cause him to avoid a number of snares which he will find in the path as he journeys on; and the second will always lend a hand to assist him, if he has unavoidably got entangled in them. The little distinctions which have been shown him at his own home ought to be forgotten when he travels over the world at large; for strangers know nothing of his former merits, and it is necessary that they should witness them before they pay him the tribute which he was wont to receive within his own doors. Thus, to be kind and affable to those we meet, to mix in their amusements, to pay a compliment or two to their manners and customs, to respect their elders, to give a little to their distressed and needy, and to feel, as it were, at home amongst them, is the sure way to enable you to pass merrily on, and to find other comforts as sweet and palatable as those which you were accustomed to partake of amongst your friends and acquaintance in your own native land.

We will now ascend in fancy on Icarian wing, and take a view of Guiana in general. See an immense plain, betwixt two of the largest rivers in the world, level as a bowling-green, save at Cayenne, and covered with trees along the coast quite to the Atlantic wave, except where the plantations make a little vacancy amongst the foliage.

Though nearly in the centre of the torrid zone, the sun's rays are not so intolerable as might be imagined, on account of the perpetual verdure and refreshing north-east breeze. See what numbers of broad and rapid rivers intersect it in their journey to the ocean, and that not a stone or a pebble is to be found on their banks, or in any part of the country, till your eye catches the hills in the interior. How beautiful and magnificent are the lakes in the heart of the forests, and how charming the forests themselves, for miles after miles on each side of the rivers! How extensive appear the savannas, or natural meadows, teeming with innumerable herds of cattle, where the Portuguese and Spaniards are settled, but desert as Sahara where the English and Dutch claim dominion. How gradually the face of the country rises! See the sand-hills all clothed in wood first emerging from the level, then hills a little higher, rugged with bold and craggy rocks, peeping out from amongst the most luxuriant timber. Then come plains, and dells, and far-extending valleys, arrayed in richest foliage; and beyond them, mountains piled on mountains, some bearing prodigious forests, others of bleak and barren aspect. Thus your eye wanders on, over scenes of varied loveliness and grandeur, till it rests on the stupendous pinnacles of the long-continued Cordilleras de los Andes, which rise in towering majesty, and command all America.

How fertile must the lowlands be, from the accumulation of fallen leaves and trees for centuries. How propitious the swamps and slimy beds of the rivers, heated by a downward sun, to the amazing growth of alligators, serpents, and innumerable insects. How inviting the forests to the feathered tribes, where you see buds, blossoms, green and ripe fruit, full-grown and fading leaves, all on the same tree. How secure the wild beasts may rove in endless mazes. Perhaps those mountains, too, which appear so bleak and naked, as if quite neglected, are, like Potosi, full of precious metals.

Let us now return the pinions we borrowed from Icarus, and prepare to bid farewell to the wilds. The time allotted to these Wanderings is drawing fast to a close. Every day for the last six months has been employed in paying close attention to natural history in the forests of Demerara. Above two hundred specimens of the finest birds have been collected, and a pretty just knowledge formed of their haunts and economy. From the time of leaving England, in March, 1816, to the present day, nothing has intervened to arrest a fine flow of health, saving a quartan ague, which did not tarry, but fled as suddenly as it appeared.

And now I take leave of thee, kind and gentle reader. The new mode of preserving birds, heretofore promised thee, shall not be forgotten. The plan is already formed in imagination, and can be penned down during the passage across the Atlantic. If the few remarks in these Wanderings shall have any weight in inciting thee to sally forth and explore the vast and well-stored regions of Demerara, I have gained my end. Adieu.

CHARLES WATERTON.

*April* 6, 1817.

# NOTES.

[24] The negroes of the West Coast of Africa, as I am informed by Dr. Kodjoe Benjamin William Kwatei-kpakpafio, of Accra, take their names from the day of the week on which they are born: Quashi (Kwasi) is Sunday; Kodjoe, Monday; Koffie, Tuesday.—N. M.

[31] "Natural History Essays," by Charles Waterton, edited, with a life of the author, by Norman Moore (Warne and Co.).

www.ingramcontent.com/pod-product-compliance
Ingram Content Group UK Ltd.
Pitfield, Milton Keynes, MK11 3LW, UK
UKHW032124030325
4838UKWH00004B/312